MEDIA AND MIXES
FOR
CONTAINER-GROWN
PLANTS

TITLES OF RELATED INTEREST

Growing media for ornamental plants and turf
K. A. Handreck and N. D. Black
University of New South Wales Press

MEDIA AND MIXES FOR CONTAINER-GROWN PLANTS

(second edition of
Modern potting composts)

A manual on the preparation
and use of growing media for pot plants

A. C. BUNT
N.D.H.(Hons), M.I. Biol.
*Formerly with the Glasshouse Research Institute,
Littlehampton, Sussex*

London
UNWIN HYMAN
Boston   Sydney   Wellington

Published by the Academic Division of
Unwin Hyman Ltd
15/17 Broadwick Street, London W1V 1FB, UK

Allen & Unwin Inc.,
8 Winchester Place, Winchester, Mass. 01890, USA

Allen & Unwin (Australia) Ltd,
8 Napier Street, North Sydney, NSW 2060, Australia

Allen & Unwin (New Zealand) Ltd in association with the
Port Nicholson Press Ltd,
60 Cambridge Terrace, Wellington, New Zealand

First published in 1976 as *Modern potting composts*
Second edition 1988

British Library Cataloguing in Publication Data

Bunt, A. C.
Media and mixes for container-grown plants:
a manual on the preparation and use of
growing media for pot plants. — 2nd ed.
1. Plants,.Potted 2. Compost
I. Title II. Bunt A. C. Modern potting composts
635.9'65 SB418
ISBN 0-04-635016-0

Library of Congress Cataloging-in-Publication Data

Bunt, A. C.
Media and mixes for container-grown plants: a manual on the
preparation and use of growing media for pot plants/A. C. Bunt. —
2nd ed. of Modern potting composts.
    p.   cm.
Bibliography: p.
Includes index.
ISBN 0-04-635016-0 (alk. paper)
1. Potting soils. 2. Plants, Potted. 3. Greenhouse management.
I. Bunt, A. C. Modern potting composts. II. Title.
S589.8.B86 1987
635.9'86—dc 19   87-15332
                                                  CIP

Typeset in 10 on 11 point Palatino by Paston Press, Loddon, Norfolk
Printed in Great Britain by Biddles of Guildford

Contents

CONTENTS

List of tables

LIST OF TABLES

xiv

Preface to the first edition

The past two decades have seen rapid advances in the technology used to produce pot plants. Glasshouses designed and orientated to give maximum light transmission, fully automatic heating and ventilating systems, carbon dioxide enrichment of the atmosphere, controlled photoperiods using automatic blackouts and incandescent lamps which enable plants such as chrysanthemum to be flowered at any time of the year, mist propagation techniques, chemical growth regulators which control the height of plants, automatic watering and feeding systems, etc.: these are only some of the developments which have transformed pot plant culture.

There have also been many changes in the composts and systems used to grow the plants. Mineral soils, which formed the basis of the John Innes composts, are now either too expensive or too difficult to obtain in suitable quality and sufficient quantity. Consequently the grower has been forced to seek other materials such as peat, perlite, vermiculite, plastic foam, shredded bark, etc. New types of fertilizers, new methods of heat sterilization and new chemical sterilizing agents are also being used.

As with many industrial processes, an alteration to improve one part of a process often means that alterations to other parts of the process are required in order to make the whole operation successful; so too with the new composts. By changing the bulky materials from which the compost is made, a different emphasis must be given to the type and quantity of base fertilizers used, and also to the watering and liquid feeding. Composts made with these new materials give results that are equal or even superior to those obtained from the traditional composts, *providing that their individual characteristics and requirements are understood*. The use of these new composts should be regarded as *a new system of growing*, rather than a simple change of compost.

The purpose of this book is to provide horticulturists, including students, growers, advisory officers and those who simply grow plants for pleasure, with information on the characteristics of the new materials, how they can be used to make composts and for the subsequent nutrition of the plants. Much of the information given is based on previously unpublished studies made by the author at the Glasshouse Crops Research Institute over the last fifteen years.

xvii

Preface to the second edition

In the decade since the first edition there has been no slackening in either the rate of annual growth in the volume of container media used, or in the number of research reports published. The numbers of plants that are grown in small volumes of media, either for the whole of the cropping period or for transplanting, continues to increase. For example, in 1984 50% of the greenhouse tomato crop in the UK was grown in peat bags or modules rather than in the border soil. Also in 1984, 76.2 million hardy ornamental nursery stock plants were grown in containers, whereas the amount in 1974 was only 19.8 million.

The economic advantages obtained from raising vegetable transplants in peat blocks, 'Speedling' or 'Hassy' type modules, and bedding plants by the plug system, have stimulated research into media and growing techniques suitable for these systems.

In western Europe peat continues to be the most important of the materials used for making growing media. There is now an increasing awareness of the need to maintain a physical balance between the water-holding capacity of media and their air-filled porosity, either by using young, fibrous, 'white' peats or by adding physical conditioners such as bark, perlite or grit where natural aeration is insufficient.

In regions where peat is unavailable or uneconomic, greater attention has been given to the use of local materials such as bagasse, coconut fibre, peanut hulls, etc. The use of shredded bark is probably the best example of research turning a waste product, that is not only potentially toxic to plants but also has special nutritional characteristics, into a suitable growing media.

Chapters have been rewritten, extended or revised, depending upon the extent of recent developments in the areas covered. The purpose of the book remains the same.

Acknowledgements

I am grateful to the many colleagues, at home and overseas, for their helpful suggestions in the preparation of this edition. In particular I wish to thank Miss M. A. Scott and Mrs C. King (ADAS) for advice on nursery stock mixes, Dr Pauline M. Smith (GCRI) for guidance on chemical sterilization and fungicides, also Mr E. W. Johnson (ADAS) and Mr P. Adams (GCRI) for helpful comment on liquid feeding. Details of potting media and practices used in other countries were supplied by Dr H. R. Gislerød (Norway), Mr D. G. Nichols (Australia) and Dr M. Prasad (New Zealand); Professor F. A. Pokorny supplied information on the pre-treatment and use of bark mixes in the USA.

I have also benefited from many useful discussions on various aspects of container media with Professor J. L. Paul and Dr D. R. Hershey at the University of California, Davis; and with Mr R. L. Jinks at GCRI. Many of the lines of research that have proved both interesting and rewarding to pursue have resulted from discussions with grower friends in the glasshouse and nursery stock industries.

Grateful acknowledgements are given to the publishers and the authors for permission to reproduce Figures 2.2 and 4.12. Figure 2.5 is reprinted by permission of Vapo Oy, and Figure 2.6 by permission of Bord Na Mona.

CHAPTER ONE

Loam or loamless media?

When plants are grown in containers their roots are restricted to a small volume, consequently the demands made on the media for water, air and nutrients are much more intense than those made by border-grown plants which have an unrestricted root-run and an infinitely greater volume of soil in which to grow. It has long been recognized by amateur and commercial growers alike that simply using a garden or border soil in a pot without any improvement to its physical properties or nutrient status will give poor results. Growers have traditionally used such materials as leafmould, decayed animal manure, spent hops, peat, mortar rubble, wood ashes, sand and grit as additives to mineral soils to improve their physical properties. The introduction of the John Innes composts by Lawrence & Newell in the 1930s did much to standardize and rationalize the multitude of materials then being used. Fifty years of experience with these composts have shown that, if prepared and used correctly, very good results can be obtained.

In horticultural terminology 'compost' has several meanings. It is used with respect to (a) plant material that has undergone biological decomposition before being dug into garden soil, (b) animal manure and straw which is used as the basis for mushroom growing and (c) mixtures of organic and mineral materials used for growing plants in containers. To avoid possible confusion to readers not familiar with its alternative usage in English, the term 'compost' has been reserved for its traditional use in describing potting media based on mineral soil, e.g. the John Innes potting composts. The term 'potting mix' has been adopted for the newer lightweight growing media, usually based on peat, bark, vermiculite, perlite, etc. 'Potting media' include both mineral soil composts and lightweight potting mixes.

The quality of the potting medium is universally recognized as being one of the foundation stones upon which the successful growing of pot plants is built. Achieving the correct physical and nutritional conditions is just as important for crops which only remain in the pots for a few weeks before being planted out, e.g. tomatoes and lettuce, as it is for plants such as *cyclamen* which are grown to maturity in pots. Any mistake made during the seedling or young plant stage is not easily rectified; often a check in growth at this stage, caused by a soil pasteurization toxicity,

1

salinity or nutritional problem, will still be evident when the plant is mature. Obviously, very careful consideration must therefore be given before making any fundamental changes to such an important part of pot plant culture; indeed, it might well be asked, 'Why change?'

There are essentially two reasons why pot-plant growers have been changing from the loam-based John Innes composts to mixes made from peat, vermiculite, plastics, etc., and other loamless materials. They are:

(a) the practical difficulty of obtaining sufficent quantities of loam which conform to the John Innes specification;
(b) economic advantages in dispensing with the mineral soil and its preparation.

To meet the requirements of the John Innes composts, the loam must be a 'medium clay', free from lime or chalk and preferably be a 'ley' or turf; this ensures that it has a good structure with sufficient organic matter or fibre to supply the plant with nitrogen over a long period. Such loams are difficult to obtain today. Legislation prevents the uncontrolled stripping and sale of agricultural soil, and soils for compost making are obtained from various sources, e.g. building sites, roadworks, etc. These soils are often quite unsuitable, they can be heavy clays which are difficult to handle, light sandy soils with poor structure, subsoils with no organic matter, or soils with a high chalk content. A further very important point to consider when selecting a soil is its reaction when pasteurized. To ensure that the loam is free from plant pathogens it is necessary that it is partially sterilized, preferably by heat treatment (Ch. 12), before it is used. Unfortunately, the beneficial effects of this treatment in eliminating the plant pathogens are accompanied by several biological and chemical changes, some of which can be detrimental to plants. One of the most important of these changes following pasteurization is the increase in the amount of water-soluble and exchangeable forms of manganese; in some soils this can reach toxic levels. Raising the pH of the soil by adding ground limestone or chalk, preferably before pasteurization, will materially reduce the amount of active manganese. Unfortunately, this treatment may also have an accompanying adverse effect with respect to the forms of nitrogen present (Ch. 5 & 12). Steam or heat pasteurization of the soil also changes the balance of the various types of bacteria which convert the organic matter into mineral nitrogen. The ammonifying bacteria are spore forming types, they are resistant to heat treatment and multiply rapidly after the soil has been pasteurized. The nitrifying bacteria, however, which convert the ammonium into nitrites and then into nitrates, are temporarily eliminated from the soil when it is heat treated; consequently one link in the normal process of nitrogen mineralization is broken and a build-up of ammonium can occur. Steam pasteurization, a high soil pH and the presence of readily decomposable organic matter are the prime factors required to create free ammonia problems (see Ch. 5 & 12). The liming of a soil to reduce manganese toxicity following steam pasteurization is a good example of remedial action for one problem possibly creating another problem.

2

Before leaving the matter of loam selection there is also the question of continuity of supply to consider. A grower with one hectare of glasshouses in a 'year-round' pot chrysanthemum programme, for example, will use about 11 m^3 (approximately 14 yd^3) of compost each week, and this will require 300 m^3 of loam each year. To maintain a week-by-week consistency of plant quality calls for an equal consistency in the quality of the compost, and securing such a quantity of good quality loam obviously presents difficulties.

Regarding the economic aspects of loam and loamless media, no critical comparisons can be made because of local differences in the basic prices of the materials and of differences in the degree of mechanization available. One very large factor in the costs of preparing media is that of heat pasteurization. Because various weed seeds and pathogens such as wire-worms and damping-off fungi, etc. are normally present in mineral soils, it is essential that they are partially sterilized before use, either by steam or flame pasteurization (Ch. 12). Such equipment is expensive both to purchase and to operate. Furthermore, the soil must be kept dry and under cover before it can be successfully treated. Not only does wet soil require considerably more steam than dry soil, it is much more difficult to treat efficiently because wet soils tend to 'puddle' and leave 'cold spots' of untreated soil. Diseases surviving in the untreated parts of the soil will then spread more rapidly into the treated soil and may cause more trouble than if the soil had not been-treated at all. By contrast, loamless mixes based on peat, vermiculite or perlite do not usually require pasteurization, and peat packed in polythene-wrapped bales can be stored in the open without the need for expensive storage facilities. Peat is not technically sterile, and it contains various micro-organisms, but these are not usually pathogenic. Consequently in Britain and western Europe it is not customary to pasteurize new peat before it is used. In the US it is often considered advisable to steam the peat; much will depend on the source of the peat and its known history of pests and diseases.

Composts made from mineral soils are also much heavier to handle, both during their preparation and in filling and moving the pots or other containers. If large quantities of loam-based compost are required, it will be more desirable to use such equipment as mechanical shovels, soil shredders, conveyor belts, etc., in order to keep down the labour costs, than if the lighter loamless mixes are used. For these reasons it is generally accepted that loamless mixes are both easier and cheaper to prepare than conventional loam-based compost.

The advantages and disadvantages of loam-based and loamless media can be summarized as follows:

1.1 LOAM COMPOSTS

1.1.1 Advantages

(a) The principal advantage is ease of plant nutrition. Composts made from a good loam have more plant nutrients and the nutrition,

3

especially with respect to nitrogen and phosphorus, is easier. Starvation occurs less rapidly.

(b) Minor element deficiencies are not common.

1.1.2 Disadvantages

(a) Difficulty in obtaining suitable loam that does not give toxicities when steam pasteurized.
(b) Continuity of supply and quality control.
(c) The loam must be stored dry and steam pasteurized before it is used.
(d) The composts are heavy and difficult to handle.
(e) They are more expensive to prepare if done properly.

1.2 LOAMLESS MIXES

1.2.1 Advantages

(a) A greater degree of standardization of materials, less variability between successive batches.
(b) Often better physical properties can be achieved by selecting appropriate grades of materials, e.g. to give a higher air-filled porosity.
(c) Do not usually require heat pasteurization.
(d) Cheaper to prepare.
(e) Lighter to handle.
(f) The initial lower nutrient content of the materials can be used to give more controlled growth.

1.2.2 Disadvantages

(a) Most of the materials have low levels of 'available' and reserve nutrients.
(b) Balanced ratios of immediately available nutrients must be added, the amounts being sufficient for good growth rates but not enough to cause salinity problems.
(c) There is a greater dependence on liquid feeding or suitable forms of controlled release fertilizers.
(d) Starvation symptoms develop more rapidly than in loam-based media.
(e) Some materials show a biological or chemical fixation of nitrogen; extra nitrogen must then be given in the base fertilizer to compensate.
(f) Microelements must also be added.

1.3 LIGHTWEIGHT MIXES WITH SOME MINERAL SOIL

1.3.1 Advantages

With a view to combining the advantages which each of the above groups of media offers, some growers incorporate a low percentage of loam in the mix. For example, such a mix may contain one-third peat, one-third of a lightweight aggregate (e.g. perlite), and one-third loam; by comparison the John Innes composts contain almost 60% loam.

1.3.2 Disadvantages

One potential disadvantage with this type of mix if the pH is kept low, e.g. at pH 5.5 as with peat mixes, is the risk of manganese toxicity occurring from the pasteurized soil. Also, unless the soil has a good physical structure there is the risk that it will reduce the air-filled porosity of the mix.

Where the amount of soil in the media is very small, i.e. 10% or less, for analytical purposes it is regarded as being a peat-based mix.

Growers must now decide which of these types of media they wish to use. A description of the materials available for making loamless mixes, together with their preparation and fertilizer requirements, is given in Chapters 2–9, and the preparations of the more conventional loam-based John Innes compost is described in Chapter 11.

CHAPTER TWO

Materials for loamless mixes

There are a number of bulky materials which can be used either separately or in various combinations to make loamless mixes, the choice of a particular material is usually determined by its availability, cost and local experience of its use. In northwest Europe peat is the material most widely used in loamless mixes, and in the USA bark, vermiculite and perlite are also extensively used.

The most important single factor when choosing a material for potting mixes is that it should be free from substances that are toxic to plants. A large number of materials fulfil this requirement and can be successfully used, providing that the management (this usually means the watering and the nutrition) is adjusted to the requirements of the media and the crop. The common conclusion that '. . . this particular mix gave the best results' really means '. . . this mix gave the best results *under the particular system of management prevailing during the experiment'*. A change in management practice or even the seasonal change in the environment can often give completely different results. The following is a brief description of the materials most commonly used for making potting mixes.

2.1 PEAT

This is by far the most widely used material for making potting mixes, either by itself or in combination with other materials. In the raw state it is usually deficient in the principal plant nutrients, but in common with mineral soils it has the important advantage of having a useful cation exchange capacity: this is a chemical mechanism which helps in regulating the supply of some nutrients to the plant (§4.1). From a casual inspection of peat in bales it might at first appear that it is a uniform and standardized product. Closer investigation however will soon show that this is not the case. Peat is formed by the partial decomposition of sphagnums, other mosses and sedges. Under acid, waterlogged conditions, and in the absence of nutrients, the micro-organisms, which would normally break down or decompose the plants, are excluded and only partial decomposition of the dead tissue occurs. Differences between peats are related to variations in local climate, the species of plants from which the peats are formed and their degree of decomposition in the bog.

PEAT AND PEATLAND TERMINOLOGY

A number of specialized terms are used to describe peat types and their formation. The following glossary has been abridged from Stanek & Worley (1984).

amorphous peat   Highly decomposed, no evident structure.
basin   A depressed area having no surface outlet.
blanket bog   A 'blanket' of peat, not more than 2 m deep, that develops in cool, temperate maritime climates.
bog   Peatlands that are dependent upon precipitation for nutrients, usually acidic and covered with *Sphagnum*.
carex peat   Containing significant amounts of sedge remains.
eutrophic   Peatlands that are relatively rich in nutrients.
fen   Often sedge-rich peatland, higher in nutrients and less acidic than a bog.
fibre   A fragment or piece of plant tissue (excluding live roots) in peat that retains recognizable cellular structure. In the USA fragments retained by a 100-mesh sieve (openings 0.15 mm in diameter).
fibric   Used in the USA and Canada to describe least decomposed peats.
hemic   Peat in an intermediate stage of decomposition, i.e. fibric → hemic → sapric.
highmoor   As for *Hochmoor*.
Hochmoor   German term meaning raised bog, mire or peatland.
humic   Used in the USA and Canada, syn. Sapric.
lowmoor   From the German *Niedermoor*, peatland receiving minerals from flowing or percolating water.
mesotrophic   Mediocre or moderate nutrient availability. Organic matter decomposition is slow.
moss peat   Syn. peat moss, sphagnum peat.
nutrient classes   The nutrient status of a site. The three most common being: poor (oligotrophic), mediocre (mesotrophic) and rich (eutrophic).
oligotrophic   Extremely low in nutrients and biological activity; organic matter little decomposed. For example raised bog of spagnum peat.
ombrotrophic   Peatland areas dependent only on nutrients from precipitation.
peat   Largely organic residues of plants, incompletely decomposed through lack of oxygen.
peatland   Areas with peat-forming vegetation growing on peat, the undrained peat being not less than 30 cm deep.
raised bog   Peatland with convex surface, peat poor in nutrients including calcium, syn. highmoor.
reed peat   Mainly reeds (*Phragmites, Scirpus, Typa*).
sapric   Highly decomposed peat.
sedge peat   Composed primarily of stalks, leaves, rhizomes and roots of sedges and *Carex* spp.
sphagnum peat   Composed mainly of weakly decomposed *sphagnum* spp. with admixed *Eriophorum* spp. and *Carex* spp.; syn. moss peat.
von Post humification scale   Decomposition scale of peat, ranging from H1 (completely unhumified) to H10 (completely humified).
woody peat   In North America, peat containing the woody remains of trees and shrubs.

Table 2.1 Peat areas and production.

Country	Area of peatland* 30 cm deep (million hectares)	Peat production† (1000 tonnes year^{-1}; 40% moisture) Horticultural	Fuel
Canada	170	400	—
USSR	150	100 000	100 000
Finland	10.4	220	1 500
USA (without Alaska)	10.25	700	—
Sweden	7	300	—
Norway	3	90	—
UK	1.58	500	50
Poland	1.35	340	—
Ireland	1.18	450	6 000
Federal German Republic	1.11	250	2 000

* From Kivinen & Pakavinem (1980).
† From Kivinen (1981).

2.1.1 Peat deposits

Although the largest peat deposits are to be found in the northern hemisphere in regions of high rainfall and low temperature, e.g. Canada, northwest Europe and the USSR, smaller deposits also occur in subtropical and tropical areas, e.g. Florida, Cuba and Indonesia. Recent estimates of the major areas of peat reserves, together with the production of peat for horticultural use and as a fuel, are given in Table 2.1. Canada has the largest area of peatland but the USSR is by far the largest producer of peat; there most of the peat burnt as fuel is used for electricity generation. Approximately one-third of the peat produced in the UK is humified sedge peat, which is largely used for tomato modules; some black humified sphagnum peats, rated H7 or H8 on the von Post scale, are also produced for making peat blocks (Ch. 14).

2.1.2 Classification of peat deposits and types

There is no universally accepted classification of peat deposits. To some extent this depends upon the purpose of the classification, whether it is made by soil scientists, botanists or horticulturists. There are also some differences in the systems adopted by various countries (Farnham 1969). The following is a general, broad classification:

Lowmoor peat is formed from the shallow flooding of a depression or the infilling of a lake. In both cases the water contains some mineral bases. Deposits of this type may be formed at any altitude but they usually occur at fairly low elevations.

Highmoor peat is formed without prior inundation, the surface being kept continuously saturated because of the moisture-holding properties of the surface as well as heavy rainfall. It usually overlies peat of the lowmoor type. Essential conditions for its formation are comparatively cold, wet conditions and a very low level of mineral bases. Bogs of this type are generally higher at the centre than at the margins and are often referred to as raised bogs. They usually contain two layers: a lower one of more humified peat and an upper one of material which is generally less decomposed; both contain considerable proportions of the remains of sphagnum moss. In Germany these bogs are known as *Hochmoor* and the peat can be 'white peat', i.e. slightly decomposed, or 'black peat', i.e. highly decomposed.

The term 'basin peat' can also be used to describe the lowmoor and highmoor types and refers to the peats formed in and over the pool or lake from which the bog was formed.

Blanket bog is similar to highmoor but is formed in areas where the surface is continuously saturated by high rainfall. It largely follows the contours of the ground and the level of mineral bases in the peat is very low. Blanket bogs generally occur on high moorland but they can also occur at sea level.

The raised bogs of the highmoor type usually give the best peats for making container media.

Several plants grow in the bogs from which the peat is formed and the peats are classified according to the principal plant species. For horticultural purposes they can be grouped under two headings: sphagnum and sedge peats.

Sphagnum. Several sphagnum mosses are found in peat bogs, the most important of which are *S. papillosum*, *S. imbricatum* and *S. magellanicum*, which are members of the *cymbifolia* group, and *S. rubellum*, *S. plumulosum* and *S. fuscum* which belong to the *acutifolia* group. Sphagnum leaves consist of a single layer of cells and those in the *cymbifolia* group have large boat-shaped leaves, this gives them a high rate of water absorption and retention. Peats from this group tend to be loose and bulky, they decompose more quickly than peats in the *acutifolia* group. Sphagnums in the *acutifolia* group have smaller leaves, retain less water and the resulting peats are denser. The general characteristics of the sphagnum peats are a spongy fibrous texture, a high porosity with high water-retaining capacity, a low ash content and usually a low pH. Some sphagnum-type peats contain variable amounts of cotton-grass (*Eriophorum*).

Sedge peat. This type of peat is formed from sedges (*Carex* spp.) and reed grass (*Phragmites*) with some Hare's tail cotton-grass (*Eriophorum* *vaginatum*) and common heather (*Calluna vulgaris*). Sedge peats have usually developed under the influence of drainage from mineral soils, consequently they contain more plant nutrients than the sphagnum peats. They are darker, more humified and decomposed than sphagnum peats and have a higher cation exchange capacity per unit weight. They have a lower water retaining capacity than sphagnum peats and although they make good agricultural land when reclaimed, they can also be used for making pot plant mixes if given the correct management. Because of their structure and better binding capacity when compressed, peats of this type are preferred to the younger, less decomposed sphagnum types for making peat blocks. These are free-standing units of fertilized, compressed peat used for propagating plants such as chrysanthemum and lettuce which are eventually planted into the glasshouse borders. The peat block is in fact somewhat similar to the older type of soil block which was made from the John Innes composts.

Classification systems. In the USA, the American Society for Testing Materials (ASTM) has proposed a system of peat classification based on the generic origin and fibre content. There are five groups in the system:

(a) *Sphagnum moss peat* (*peat moss*). An oven-dry peat sample contains over 75% by weight of sphagnum moss fibre. The fibre should be stems and leaves of *Sphagnum* spp. in which the fibrous and cellular structure is recognizable. The samples must contain a minimum of 90% of organic matter on a dry weight basis.

(b) *Hypnum moss peat*. The hypnum moss fibre content must be more than 50% of the oven-dry weight of the sample, and the organic matter content must be not less than 90% of the oven-dry weight. The fibres should be stems and leaves from plants of various *Hypnum* spp.

(c) *Reed-sedge peat*. An oven-dry sample of the peat must contain a minimum of 33.3% by weight of reed, sedge or grass fibres, i.e. non-moss fibres.

(d) *Peat humus*. The total fibre content of the oven-dry peat contains less than 33.3% by weight of total fibre.

(e) *Other peat*. Peats which are not classified by the previous groups.

For the purpose of this classification the term fibre is described as plant material retained on a 100-mesh sieve or larger. This includes stems, leaves or fragments of bog plants. Wood particles larger than 1.3 cm (0.5 in.) are excluded and also mineral matter such as stones, gravel, shells, etc.

The International Peat Society has proposed a simplified 3 × 3 classification system for peat, based on its botanical composition, degree of decomposition and tropic (nutrient) status (Kivinen 1980).

(a) Botanical composition:

    (i)   moss peat (predominately sphagnums and other mosses),
    (ii) sedge peat (sedges, grasses, herbs and related),
    (iii) wood peat (the remains of trees and woody shrubs).

(b) Degree of decomposition:

    (i)   weakly decomposed ($\equiv$ fibric, von Post H1–H3),
    (ii) medium decomposed ($\equiv$ mesic, H4–H6),
    (iii) strongly decomposed ($\equiv$ sapric, H7–H10).

(c) Trophic status:

    (i)   oligotrophic (low in base salts that are plant nutrients; e.g. sphagnum and calluna peats),
    (ii) mesotrophic,
    (iii) eutrophic (high in salts that are plant nutrients; e.g. sedge, reed and fen peats).

These categories refer mainly to the pH, calcium and magnesium contents and degree of base saturation.

Horticultural classification. In commercial horticulture, 'sedge' peat is often used to describe peat which is dark, humified and having a poor structure. It may contain decomposed sphagnums as well as sedges. 'Sphagnum' or 'moss' peat is used to describe the less humified peats derived from the upper layer of raised bogs. The following general classes are also used:

Class	Degree of decomposition (von Post)
light peat	H1–H3
dark peat	H4–H6
black peat	H7–H10

2.1.3   Peat texture and structure

As peats decompose the particles become smaller, this reduces the total porespace, but more significantly it reduces the volume of large air-filled pores. The texture or particle size is also dependent upon the species of moss and the method of harvesting; milling can produce very fine particles, if the peat is humified and not fibric, resulting in poor aeration when the peat is in shallow containers. Peats are often described as coarse, medium or fine texture, and these grades have been defined in the USA and Norway (Table 2.2).

Table 2.2 Peat textural grades and particle sizes.

Grade	ASTM Particle sizes (mm)	Norwegian standard (NS 28090E)	
		All particles under:	Particles >1 mm (%)
coarse	>2.38	40 mm	70
medium	2.38–0.84	15 mm	60
fine	<0.84	6 mm	30

Sieving is a quick and easy way of grading peat, but the results depend upon the moisture content when it is sieved. Air drying may result in the fragmentation of young sphagnum peats during sieving (Boggie & Robertson 1972) while more humified sphagnum and sedge peats can form stable aggregates on drying. Desorption curves are more slowly obtained but they provide the best assessment of peat texture; curves for a range of peats are given in Figure 2.1.

Young sphagnum peats with very fine particles can form weak aggregates when the dry particles are moistened. Apart from a shallow layer of peat just below the container surface, where the aggregates are dispersed by the turbulence during irrigation, these aggregates are sufficiently stable to help create a good structure.

The degree of peat decomposition can be measured in several ways. Two of the best known are the von Post and the volume weight methods. The von Post method consists of a scale of 10 grades designated H1 to H10. It is based on assessing the quality of the water which is exuded when wet peat is compressed in the hand, and also on the amount of peat

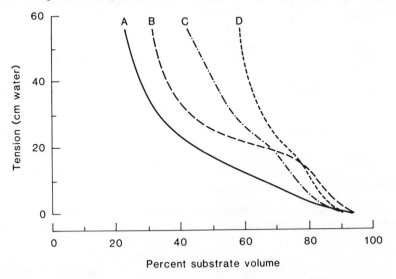

Figure 2.1 Desorption or water release curves of four peats with differing structures. *Key*: A young fibrous sphagnum, coarse particles; B young sphagnum, fine particles; C sphagnum, medium decomposition; D sedge peat, highly decomposed.

Table 2.3 The von Post scale for measuring peat decomposition.

Degree of decomposition	Quality of water exuded	Proportion of peat pressed out between fingers
1	clear, colourless	none
2	almost clear, yellowish brown	none
3	turbid, brown	none
4	very muddy brown	none
5	extremely muddy, contains little peat in suspension	very little
6	dark, plenty of peat in suspension	one third
7	very dark, thick	half
8	very thick	two thirds
9	no free water	nearly all
10	no free water	all

which is pressed out between the fingers (Table 2.3). This test is sometimes supplemented by determining the volume weight or bulk density of the peat, i.e. the weight of peat per unit volume. A standard procedure is used to wet the peat and allow it to settle to a natural density without any compression. Puustjärvi (1970b) has shown that a good correlation exists between the degree of decomposition as determined by the von Post scale and the bulk density (Fig. 2.2).

He also found (1983) a linear relationship between the exchange capacity of peat and its natural bulk density (BD). The least decomposed sphagnum peat, H1 on the von Post scale with a BD of $45\,g\,l^{-1}$, had a cation

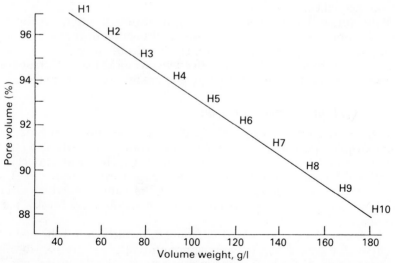

Figure 2.2 Decomposition of peat. The relationship of the pore volume, the volume weight or bulk density and the von Post scale of decomposition (from *Peat and Plant News* 1970).

13

exchange capacity of 100 meq 100 g^{-1} or 45 meq l^{-1}. At H3 the BD had increased to 75 g l^{-1} and the exchange capacity to 120 meq 100 g^{-1} or 90 meq l^{-1}; at H5 (the maximum desirable for peat in a potting mix) the BD was 105 g l^{-1} and the exchange capacity 124 meq 100 g^{-1} or 130 meq l^{-1}. Black peat with a decomposition of H8 had an exchange capacity of 116 meq 100 g^{-1} or 174 meq l^{-1}.

Another method of assessing the humification or degree of decomposition of peat is by hydrolysis of the dry organic matter with sulphuric acid (Federal German Republic Standard DIN 11542, 1978). The amount of non-hydrolysable material (known as the r value) increases during the formation of the peat and can be used to measure the degree of humification. These two methods of assessing decomposition are based on different principles of measurement; an approximate relationship can be obtained from:

$$(\text{von Post}) \, H = 0.155 \, r - 2.9$$

Black peats Beneath the top layer of young sphagnum moss in a raised bog there will usually be older and more decomposed peat known as 'dark' or 'black peat', H6–H7 on the von Post scale. In the freshly harvested condition these peats have poor physical properties for use in potting mixes; howeyer, they have greater cohesion than young sphagnum peats and are used to make peat blocks (Ch. 14). Such peats show an irreversible loss of water after air drying and have poor aeration. If the fresh peat is frozen the pores are enlarged, the water released at tensions below 100 cm is increased and aeration improved; there is also less shrinkage on drying and the irreversible loss of water is less (van Dijk & Boekel 1965). The temperature of the peat must be below $-5\,°C$ for at least 3 days, and to obtain the maximum effect the moisture content before freezing must be high.

If the climate does not permit freezing of the peat, its physical properties can be improved by mixing with an equal volume of coarse undecomposed sphagnum peat (Verdure 1981). Vermiculite of 'micron' grade (# no. 4 USA) at the same rate is also effective but blocks made from peat and vermiculite have less tensile strength and are more easily damaged.

2.1.4 Peat types and plant response

As well as responding to physical differences between peat types, e.g. their air-filled porosities or available water contents, plants also react to certain of the chemical characteristics of peats. The most important of these will be the amount of available nitrogen in the peat (§5.1), the amounts of microelements present and the extent to which drench applications of chemicals, such as growth regulators, are inactivated by the peat.

The latter effect is illustrated in Figure 2.3 which shows the results of applying the growth regulator phosphonium chloride (Phosfon) to young chrysanthemum plants growing in mixes made with 25% sand and 75% of either sedge, Irish sphagnum or Finnish sphagnum peats. Optimal

14

Table 2.4 Composition of three peats (percentage of total organic matter).

Peat type	Hemi-celluloses	Cellulose	Lignin	Protein
sphagnum/cotton grass	16.6	14.0	32.8	3.6
scirpus type	22.4	13.8	41.4	4.1
phragmites types	3.4	8.4	50.8	6.4

rates of Phosfon application varied with the peat type. With the sedge peat mix 0.75 g l^{-1} was not sufficient to obtain the desired reduction in height. With the Irish sphagnum mix 0.5 g l^{-1} was optimal, whereas for the Finnish peat mix 0.25 g l^{-1} or even slightly less was sufficient to meet some market preferences. A similar effect has been obtained using chlormequat chloride (Cycocel) drenches for poinsettia height control. Plant response also varies with the season, less growth regulator being required to give the same effect in winter as in summer. When drenches of chemicals are used with a new peat type a small trial should first be made. However, there will be very little difference in plant response between peat types when using foliar spray applications.

Differences between the responses in young, less humified sphagnum peats and the sedge peat is believed to be related to their cellulose and lignin contents. Ogg (1939) found large differences in the composition of three peat types (Table 2.4).

2.1.5 Volume measurement of peat

Loose peat is easily compressed thereby making accurate volume measurements difficult. Weight can be determined more accurately, but as peats absorb large amounts of water even greater errors would occur if no correction for moisture content was made. A standard procedure for volume measurement is therefore desirable for three reasons: (a) in order to know the bulk volume when peat is purchased loose, (b) for the preparation of potting mixes and (c) to allow chemical analysis on a volume sample (§ 4.4). Standard procedures for making volume measurements of peat are given in the Federal German Republic Standard DIN 11542 and the Norwegian Standard NS 2891.

In the UK small quantities of peat are measured by a FIBSPAN litre, a method developed by Fisons after the French Afnor Normes. The peat is lightly compressed by applying a 650 g weight, equivalent to 8.6 g cm^{-2}. For larger volumes, samples of 50 l are measured in a box 50 cm wide × 40 cm deep × 25 cm high, without compression. Tests have shown the two methods give similar results, the self compression of the peat in the large measure being similar to the applied compression with the FIBSPAN method.

The two main factors affecting the accuracy of volume measurements are: (a) whether the peat has been fully loosened after being compressed in a bale, and (b) subsequent expansion if the peat is moistened.

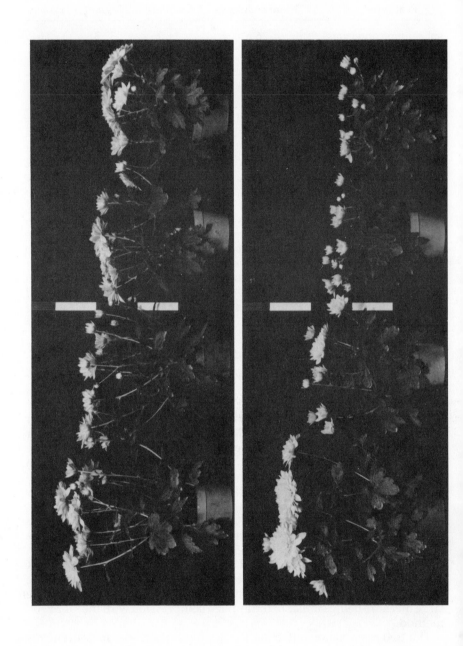

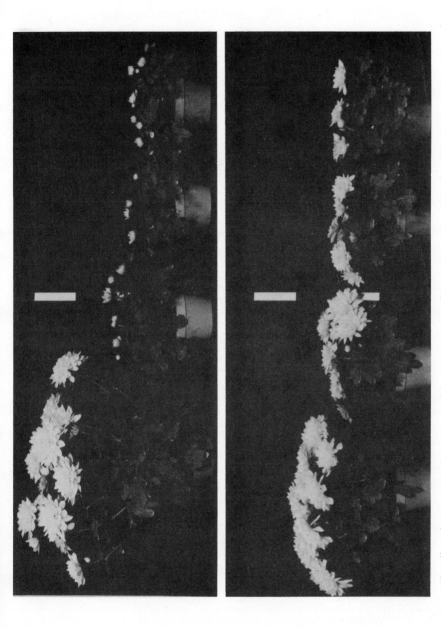

Figure 2.3 The effect of peat type on the response of chrysanthemum to the growth regulator Phosfon added to the medium. Top to bottom: sedge, Irish and Finnish peat mixes and John Innes compost. Left to right: control plants, Phosfon at 0.25, 0.50 and 0.75 g l^{-1}.

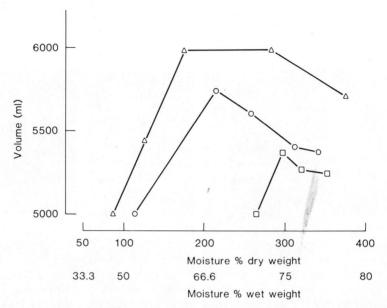

Figure 2.4 Moisture-related changes in the volume measurement of peat. The increase in volume with moisture content is more dependent upon the initial moisture content of the peat than its degree of decomposition. Sedge peats also show a large increase in volume if they are relatively dry before moistening.
Key: △ young fibrous sphagnum; ○ sphagnum, medium decomposition; □ sedge peat.

The extent to which peats expand when moistened to 'potting condition' depends upon the type of peat and the amount of water required to reach this condition. Bales of young sphagnum peats from Scandinavia usually have low moisture contents, about 90% on a dry weight basis or 47% of the moist weight. These peats when loosened, measured and then wetted can increase in volume by 20–25% (Fig. 2.4); at very high water contents the volume will decrease, especially if the peats are fibrous. Other sphagnum and sedge peats usually show smaller increases in volume on moistening, largely because they have a higher moisture content when purchased. If they are dried, measured and then rewetted they also show a large increase in volume.

A general summary of the most important physical and chemical characteristics of sphagnum peat is given in Table 2.5.

Particles should grade up to 10 mm for granular peats, 15 mm for medium, fibrous peats and 40 mm for coarse, fibrous peats; less than 40% by weight should be below 1 mm.

2.1.6 Peat production

In recent years there have been considerable changes in the methods of peat production; it is now a highly mechanized process employing large

Table 2.5 Properties of sphagnum peat.

bulk density $(g\,l^{-1})$	60–100
pore volume (%)	>96
organic matter (%)	>98
ash (%)	<2
total nitrogen (% weight)	0.5–2.5
cation exchange capacity $(meq\,100\,g^{-1})$	110–130
pH (in water)	3.5–4.0
bales	
northwest Europe, volume (l)	300–350
northwest Europe, weight (kg)	50–55
USA, volume (l)	113–155
USA, volume (ft^3)	4–5.5

and expensive equipment. A peat bog can occupy an area of several hundred hectares and it is prepared for harvesting by first cutting a series of connecting drainage trenches. After being drained, the surface layer of moss and heather is removed and production can then commence. Three distinct methods are used to harvest the peat from the bog: they are known as milling, sod cutting and hydraulic mining. The method chosen depends on the nature of the bog, whether it contains undecomposed tree trunks etc., and the local climate. With the milling process, large rotary-cultivator type machines are used to loosen the top 15 mm ($\frac{1}{2}$ in) layer of peat (Fig. 2.5).

Figure 2.5 Harvesting a young sphagnum by milling (by permission of Vapo Oy).

19

This is allowed to dry before other machines form it into ridges and load it ready for transporting off the bog. This method of harvesting can only be used when the surface layer of the peat bog has dried out sufficiently to allow the peat to be milled; it is not a suitable method in areas where dry periods are short and infrequent.

With the sod system of harvesting, brick-shaped units of peat are cut from the vertical side of a trench by a machine which also lays the peat sods onto the surface of the bog to dry (Fig. 2.6). After the initial drying, the sods are mechanically gathered to form loose hedges or ricks. With this system of harvesting, the peat is usually left on the bog to be subjected to a period of low temperature before it is taken to a factory for processing.

Hydraulic mining is practised on some bogs in Vancouver which have large amounts of tree stumps and roots. In one system peat is washed from the edge of large ditches with jets of water, the peat slurry is then collected and pumped to a dewatering plant. With another system of hydraulic mining, an excavator with a clamshell or mechanical grab is mounted on a hover barge. Peat is moved from the bog to the barge and separated from the roots by washing through a screen. The slurry is collected in a tank and pumped to a dewatering plant (Carncross 1984).

A light railway may be used to transport the peat from the bog to a factory for further processing, i.e. shredding (if necessary), sieving, grading and finally compression into polythene-wrapped bales. When most of the peat has been removed, the bog is usually brought into cultivation with vegetable or agricultural crops. It is then referred to as a 'cut-over' bog.

Figure 2.6 Peat harvesting with a sod-cutting machine (by permission of Bord Na Mona).

2.2 BARK

Bark was formerly regarded as a waste product and disposed of by burning, now its potential as a component of potting mixes is accepted. In some instances its primary function is to improve the aeration of the mix, in others it replaces peat for reasons of economy or difficulty of supply. Used correctly it gives excellent results, however it is a naturally variable material and often cannot be safely used in the fresh state. Experience has shown that variability can be due to the type of wood from which it is obtained (i.e. hardwoods or softwoods), the species of tree and its age, and the soil type and region where it was grown. Two major problems that occur in varying degrees are:

(a) nitrogen deficiency, induced either by the biological decomposition of the large amounts of readily degradable carbon (cellulose), or the chemical and physical fixation of nitrogen within the bark particles; and
(b) the presence of toxins of organic or inorganic origin, e.g. phenols and manganese.

Bark is a generic term which includes the inner bark, or living phloem, and the outer bark, or rhytidome, which includes the cork cells, or phellem. The bark usually forms 6–16% of the volume of a tree, 10% being an average value; with *Melaleuca leucandendron*, however, it is approximately 50% of the tree volume (Poole & Conover 1979). Bark is removed from the logs by drum or ring debarkers, e.g. Rosserhead type. If the bark is in large strips it must be hammer milled before being composted or used in potting mixes. The characteristics of softwood and hardwood bark are sufficiently distinct to consider them separately.

2.2.1 Softwood bark (SWB)

The species of tree most commonly grown varies with the country and region. In the USA the term 'Southern pines' includes *Pinus taeda* (loblolly pine), which is the most common of the pines, *P. echinata* (shortleaf pine), *P. elliottii* (slash pine) and *P. palustris* (long leaf pine) (Wilkerson 1981). In Norway *Picea abies* (spruce) is the predominant species (Solbraa 1979a); in Australia the most widely grown are *Pinus radiata* (Monterey pine) and *P. pinaster* (maritime pine); and in the UK *Picea sitchensis* (Sitka spruce) predominates, with *Pinus sylvestris* (Scots pine) and *P. nigra* var. *maritma* (Corsican pine) also grown.

Physical characteristics The balance between water retention and aeration in a potting mix is determined by the size distribution of the particles and pores (Ch. 3). For a mix based only on milled pine bark, Pokorny (1979) recommends a particle size distribution as in

Table 2.6 Recommended particle size distribution for pine bark and hardwood bark.

Pine bark		Hardwood bark	
Size (mm)	% by weight	Size (mm)	% by weight
>4.76	0.4	>6.4	0.7
2.38	18.6	3.2	12.9
2.00	9.0	1.6	32.6
1.00	28.3	0.8	17.6
0.84	5.4	0.5	9.7
0.59	11.3	<0.5	26.5
0.42	8.5		
<0.42	18.6		

Table 2.6. The bulk density of the milled pine bark ranges from $0.25 \, g \, cm^{-3}$ for coarse particles to $0.45 \, g \, cm^{-3}$ for fine particles (Brown & Pokorny 1975). Structural examination of milled pine bark particles (Airhart *et al.* 1978a) showed that there are openings of 5–60 μm diameter with interconnecting channels which allow water and nutrient storage within the particles. Subsequent work (Pokorny & Wetzstein 1984) has shown that milled pine bark particles have an internal porosity of about 43%. Some of the water held within the particle is available to plants provided that the roots penetrate into the particle.

Chemical characteristics The mineral composition of SWB varies with the tree species and the soil type upon which it is grown; Table 2.7 gives the analyses of a sphagnum peat and the barks from two species of softwood trees (Solbraa 1974). Both barks are much higher in phosphorus, potassium and calcium than the sphagnum peat. The minor element composition (p.p.m.) of pine bark reported by Pokorny (1979) is typical:

Boron	Copper	Iron	Manganese	Zinc
9.3	76.9	790	118	111

Sometimes the manganese levels are much higher, e.g. Solbraa & Selmer-Olsen (1981) reported concentrations of between 480 and 1070 p.p.m.

Table 2.7 Element composition of softwood barks and a sphagnum peat.

	Element composition (mg l^{-1})				
	nitrogen	phosphorus	potassium	calcium	magnesium
bark from *Pinus sylvestris* (Scots pine)	310	25	120	395	25
bark from *Picea abies* (spruce)	440	70	340	1200	110
sphagnum peat	450	2	14	150	20

Problems with manganese toxicity and methods of control are discussed in Sections 2.2.3 and 2.2.4.

Frequently plants grown in mixes having a high proportion of SWB show some degree of nitrogen deficiency unless extra nitrogen is added, in either the base fertilizer or the liquid feed. Although the C:N ratio of softwood barks is high in relation to some other organic materials, it can be shown that there is no simple relationship between this and the occurrence of nitrogen deficiency. For example, Allison & Murphy (1963) examined the decomposition rates of several pine barks when mixed with soil. With lodgepole pine, having a C:N of 275, 24.4% of the carbon was oxidized in 53 days; whereas with sugar pine, having a C:N of 311, the amount of carbon oxidized in the same time was only 3.6%. Thus two barks having approximately similar C:N ratios had very different rates of decomposition. Other work also shows wide variation in the decomposition rates of softwood barks, e.g. 17.6% of the carbon in cedar bark oxidized in 53 days and only 2.2% of the carbon in redwood bark oxidized in 63 days. The wood of the tree usually oxidizes more quickly than the bark (Table 2.8), and if much wood is removed in debarking the risk of nitrogen deficiency will be greater. In trials carried out at Efford Experimental Horticulture Station using container-grown hardy ornamentals, Scott (1984) concluded that as well as supplying the normal slow-release fertilizers (Osmocote or Ficote) to the mix, additional nitrogen as ammonium nitrate was required at a rate of 100 g m^{-3} for each 10% of uncomposted pine bark in the mix. Pokorny (1979) reported that a lower nitrogen requirement, 148 g nitrogen m^{-3} (0.25 lb yd^{-3}), was sufficient to provide adequate nitrogen for the micro-organisms acting on the pine bark; this would be equivalent to ammonium nitrate at a rate of 424 g m^{-3} (0.95 lb yd^{-3}).

As well as the reduction in mineral nitrogen by biological immobilization, there are two other ways in which the availability of nitrogen is reduced. Foster et al. (1983) found that columns of milled pine bark irrigated with an ammonium nitrate solution adsorbed 1.5 mg of ammonium nitrogen per gram of bark. This is equivalent to 1500 mg for a nursery container of three litres capacity, and it was recommended that sufficient preplanting ammonium nitrogen be added to satisfy the binding sites in the bark. It has already been established that pine bark particles

Table 2.8 Average percentage of carbon released as carbon dioxide from woods and barks over 60 days.

Type of wood	Wood carbon released as carbon dioxide		Bark carbon released as carbon dioxide	
	no nitrogen	nitrogen added	no nitrogen	nitrogen added
softwood	12.8	12.0	8.8	8.2
hardwood	30.3	45.1	22.4	24.5

The comparative value for wheat straw was 54.6%.

have an internal porosity of about 43%. When air-dry particles were submerged under reduced pressure in a solution containing 200 p.p.m. nitrate nitrogen, less than 40% of the nitrogen was recovered by normal extraction techniques (Pokorny *et al.* 1977). However, some of the retained nitrogen would ultimately be available to the plant by slow diffusion and as its roots grow into the particles.

Pine bark contains more phosphorus than sphagnum peat (Table 2.7) but much of this is water soluble and easily leached. Only small amounts of phosphorus are bound by the bark, and a concentration of 10 p.p.m. phosphorus in the medium extract solution was sufficient for the maximum growth of 'Helleri' holly (Yeager & Wright 1982).

The pH of fresh pine bark in the USA is about 4.0–4.3, but in the UK samples usually have a pH in the range 5.0–5.2; if it is composted the pH increases to 6.0–7.0. The cation exchange capacity (CEC) varies with the pH and the tree species, uncomposted bark has an average CEC of 11–13 meq 100 cm^{-3} (equivalent to 52–57 m eq 100 g^{-1}) (Brown & Pokorny 1975).

Pine bark can also inactivate some plant growth regulators. When ancymidol was applied as a media drench it was less effective in reducing the height of chrysanthemums if they were grown in pine bark mixes than if they were grown in a silty clay loam and sand mix (Tschabold *et al.* 1975).

2.2.2 Hardwood bark (HWB)

In some areas, e.g. the Midwest and Northern States of the USA, pine bark is not readily available and HWB is used instead. The main differences between the two groups are:

(a) HWB contains up to 40% cellulose which degrades quickly and causes nitrogen deficiency, whereas SWB has only 5% cellulose and therefore the biological requirement for nitrogen is less;
(b) most HWB is phytotoxic if used in the fresh state, whereas some, but not all, SWB can be used without composting; and
(c) HWB is more suppressive of several root-infecting fungi and nematodes than SWB.

Physical characteristics Recommended particle size distributions for HWB (Klett *et al.* 1972) are similar to those for SWB (see Table 2.6). After passing the bark through a 12.8 mm screen, between 20% and 40% of the particles (by weight) should be less than size 0.8 mm. Composted HWB continues to degrade, but at a slower rate, after it is used for potting; some physical changes in HWB mixes will therefore occur during the growing period, principally an increase in water retention and a reduction in the air-filled porosity. Composted hardwood bark is not as hydrophobic as fresh pine bark.

Chemical characteristics HWB decomposes at about three times the rate of SWB (see Table 2.8), consequently there would be a large check to plant growth in HWB mixes without additional nitrogen. As almost all HWB

24

releases phytotoxic compounds during degradation, and composting is therefore mandatory, the extra nitrogen required is added before the bark is composted. This ensures more rapid composting and also avoids toxicities to plants before the mineral nitrogen has been converted to the organic form.

The pH of HWB is usually higher than that of SWB and HWB also has much more calcium (i.e. about 4%, compared with 0.4% in pine bark). During composting the pH of HWB rises above 7.0 and, to avoid probable nutritional disorders, the addition of either sulphur at a rate of 1.2 kg m^{-3} of bark (2 lb yd^{-3}) or sulphur at half this rate plus an equal amount of iron sulphate is recommended (Gartner *et al.* 1973).

2.2.3 *Phytotoxicity*

When used in the fresh state many barks are phytotoxic to young seedlings. The degree of toxicity depends upon several factors, including the age of the bark and the season in which it was removed, the species of tree and the region where it had grown. Bark from old trees (35 years old) is more toxic than bark from young trees (15 years old), with the lower bark of the tree being more toxic than the upper bark. Also, bark removed in winter is more toxic than that removed in spring. In the Southern USA, fresh barks of *Pinus taeda*, *P. echinata* and *P. strobus* are not toxic; however, bark from *P. taeda* grown in South Africa is toxic (Pokorny pers. comm.). The bark of *P. elliottii* is reported as being non-toxic in the Southern USA (Cobb & Keever 1984), but is phytotoxic in Australia (Nichols 1981). The phytotoxicity of bark to transplants decreases as the size and age of the transplant is increased. Also the larger the amount of non-toxic medium retained around the roots during transplanting, the lower toxicity will be.

Identification of the toxins present in softwood barks has been studied by Yazaki & Nichols (1978), Solbraa (1979b), Solbraa & Selmer-Olsen (1981), and Aaron (1982); toxins in hardwood bark were identified by Still *et al.* (1976). The toxins can be of organic or inorganic origin.

Organic compounds Yazaki & Nichols (1978) identified a number of phenolic compounds in *P. radiata* bark. The degree of toxicity was related to the concentrations of catechin and 3,5,3',4'-tetrahydroxystilbene leached from the bark. Solbraa (1979b) found that while root growth could be stimulated by small concentrations of the phenol, pyrogallol, at high concentrations growth was reduced and finally eliminated. Concentrations of biologically active compounds were greatest in the living bark (phloem) and lowest in the dead outer bark (rhytidome). The main effect was attributed to tannins, which constituted 12% of the fresh spruce bark. Still *et al.* (1976) examined the effects of bark extracts and a number of phenolic compounds on the root growth of cucumber and mung bean. Extracts from the bark of silver maple (*Acer saccharinum* L.) were particularly toxic, as was tannic acid; chromatography and spectral analysis of the solutions showed similar patterns. Aaron (1982) concluded that the growth inhibition of plants grown in bark was related to the monoterpene content of the bark. The bark from Norway spruce and Sitka spruce, each of which caused severe inhibition of growth, had 0.3% monoterpenes, whereas bark from Scots pine, Corsican pine and Japanese larch, all of which had little or no effect on growth, had less than 0.03% monoterpenes.

Inorganic toxins The principal inorganic toxin found in fresh bark is manganese. Morris & Milbocker (1972) concluded that the depression in growth and the development of chlorotic leaves of Japanese holly grown in a hardwood bark mix was caused by manganese toxicity. Extracts of plants grown in bark had a concentration of 1125 p.p.m. manganese, compared with a concentration of 588 p.p.m. manganese in extracts of plants grown in sand. Solbraa & Selmer-Olsen (1981) found the manganese content of the bark of Norway spruce (*Picea abies*) to vary from 480–1070 p.p.m., with the exchangeable fraction being between 35 and 85%, i.e. 170–660 p.p.m. To reduce the manganese to an acceptable level the bark should be composted before it is used. The Norwegian Standard for bark (NS 2897) states that the manganese content, measured after extracting 50 g of moist bark with 100 ml of 0.5 M magnesium nitrate for 1 h, should not exceed 0.02% of the dry matter. Also, the chlorine content, which could be high if the logs have been in salt water, should not be above 0.10% of the dry matter. To overcome the unfavourable Fe:Mn ratio, which often results in symptoms of iron deficiency and manganese toxicity in plants grown in Sitka spruce bark, Wilson (personal communication) recommends the addition of either ferrous sulphate or chelated iron at a rate of 50 g m^{-3} of bark mix.

2.2.4 Ageing and composting bark

The organic and inorganic toxins in bark can be eliminated or neutralized by ageing or composting the bark before it is used. Composting has additional advantages, including that of increasing the wettability of the pine bark.

Ageing Often pine bark is hammer milled and stockpiled before it is used in growing media; this may be for reasons of convenience or as a deliberate policy of ageing. With several species, ageing is sufficient to eliminate any toxicity. If milled bark from 15-year-old *P. radiata* trees was kept moist for 90 days, the check to plant growth was largely eliminated. However, when this procedure was carried out on the lower bark from 35-year-old trees, the toxicity was reduced but not eliminated (Yazaki & Nichols 1978). When milled pine bark is stockpiled in the open the temperature of the heap rises, and leaching by rainfall also occurs; both these processes are known to reduce the amount of phenols in the bark. The efficiency of the ageing process will be increased if the piles are turned at intervals and watered when necessary.

Composting This term is reserved for the process of biological decomposition of organic matter under controlled conditions. It usually involves the addition of nitrogen and possibly other fertilizers. The composting of bark has been reviewed by Hoitink & Poole (1980).

The composting process has three phases: in the initial phase, lasting 1–2 days, the easily degradable soluble compounds are decomposed; in the second, or thermophyllic stage, the temperature rises to 40–80 °C and cellulose is degraded; in the third phase the temperature declines and the bark is recolonized by various organisms. There are distinct changes in the populations of bacteria and fungi during these stages. The duration of the composting period is determined by the type of organic matter, the temperature, moisture content, pH and the amount of available nitrogen. The temperature in the pile should rise to 40 °C, but may exceed 70 °C for a period, and the water content should be 50–70% of the moist weight. Excessive wetness, very small particles of bark and large stockpiles can cause the oxygen content to fall below the desired level of 5–12%, this can be corrected by turning the pile. If the pH of the pile is low, urea is the preferred form of nitrogen, partly because it raises the pH, and also because ammonium nitrogen is a good substrate for the micro-organisms. Composting improves the wettability of pine bark but bark should not be milled after composting as this exposes fresh water-repellent surfaces, thus decreasing the wettability, as well as exposing cellulose and tannins that have not been degraded.

Hardwood barks contain more cellulose than softwood barks and therefore require more nitrogen when decomposing. For every 50 parts of carbon assimilated by the microbes that decompose bark or sawdust, there is a requirement for 1 part of nitrogen, 0.5 parts of phosphorus and 0.1 part of sulphur.

Gartner and co-workers in Illinois (1973) recommended the following fertilizer additions to a mix of 2/3 by volume ground bark and 1/3 coarse sand: ammonium nitrate 3.56 kg m^{-3} (6 lb yd^{-3}), superphosphate 2.97 kg m^{-3} (5 lb yd^{-3}), sulphur 0.59 kg m^{-3} (1 lb yd^{-3}) and iron sulphate 0.59 kg m^{-3} (1 lb yd^{-3}).

The amount of ammonium nitrate used is equivalent to 1.9 kg nitrogen per m^3 of bark. Plant growth was inferior in mixes where the bark was composted with urea, ammonium sulphate or sodium nitrate as the source of nitrogen. Large amounts of ammonium-producing fertilizers, such as urea, could cause toxicity from free ammonia at the high pH values usually found in composted hardwood bark. Because the final pH of composted HWB is high, between 7.0 and 8.0, no lime is required for composting and sulphur is added to reduce the pH of the mix. The iron sulphate improves the Fe:Mn ratio of the bark is well as helping to lower the pH.

The fertilizers are thoroughly mixed with the bark in a rotary mixer and then stockpiled for 30 days, during which time the moisture content must be kept above 60%. No further fertilizer is required before the mix is used but the plants are fed with a 20–20–20 liquid feed of 250 p.p.m. nitrogen, 110 p.p.m. phosphorus, 207 p.p.m. potassium.

Softwood bark requires less nitrogen than HWB. Solbraa (1979c) recommends the addition of urea at 2 kg m^{-3} (3.4 lb yd^{-3}) and superphosphate at 1.5–2.0 kg m^{-3} (2.5–3.4 lb yd^{-3}). This is equivalent to 0.92 kg nitrogen m^{-3}, which is about half the rate recommended by Gartner *et al.* (1973) for HWB. Usually lime is not required for composting SWB, but where the manganese levels are high, as in some spruce barks, Solbraa & Selmer-Olsen (1981) recommend the addition of 6 kg m^{-3} of Dolomite lime.

Composting reduces the physical structure of the bark, which can then resemble that of a woody peat. For this reason aged bark is usually regarded as a better physical conditioner for mixers than composted bark.

2.2.5 *Suppressive effect of bark*

One of the beneficial effects of including composted HWB in container mixes is its suppressive action on various soil fungi. Composted HWB is known to suppress *Phytophthora*, *Pythium* and *Thielaviopsis* root rots, *Rhizoctonia* damping off, *Fusarium* wilt and some nematode diseases. The bark should form more than 50% of the mix and not contain any wood. The suppressive action is largely due to antagonistic organisms which colonize the bark after composting, but compounds with natural fungicidal properties have also been identified. Composted SWB does not have such a strong suppressive action as HWB, it suppresses *Pythium* and *Phytophthora* but not *Rhizoctonia* (Hoitink 1980).

2.3 OTHER ORGANIC MATERIALS

Although peat and bark are the most important organic components of mixes, numerous other materials are also used; their selection is largely dependent upon local availability and costs. The range of these alternative materials is much too wide to discuss their individual characteristics, the following are examples to show the diversity: aquatic plant compost (Lumis 1980), peanut hulls (Johnson & Bilderback 1981), rice hulls and cocofibre (Verdonck *et al.* 1983) and sugarcane bagasse (Higaki & Poole 1978). Such materials can be improved physically by blending with peat, perlite, etc., and, where necessary, by leaching to remove excess salts or by composting to improve C:N ratios and eliminate toxins. The following organic materials are more generally available.

2.3.1 *Sawdust*

The wood of trees decomposes more rapidly than the bark (see Table 2.8). To prevent acute nitrogen deficiency and possible phytotoxicity, the sawdust must first be composted. This will not, however, prevent phytotoxicity if certain timber preservatives have been used, e.g. boron. Softwoods immobilize much less nitrogen than hardwoods. For example,

Allison *et al.* (1963) found that the amount of nitrogen immobilized by softwoods in 160 days was equivalent to 0.59% of the weight of the wood, the corresponding figure for hardwoods was 1.10%, i.e. almost double. Shortleaf pine, however, had a much higher nitrogen requirement than other softwoods. Worrall (1981a) composted *Eucalyptus* sawdust with 2.6 kg m^{-3} urea, 1.5 kg m^{-3} superphosphate, 1 kg m^{-3} potassium sulphate, 9 kg m^{-3} Dolomite lime and a trace element mixture for 6 weeks; he concluded that sawdust could be used as a substitute for at least 50% of the peat in a mix. Composted sawdust has lower levels of calcium, nitrogen and phosphorus than bark and should always be checked for free ammonia before it is used. Although steaming has decreased the toxicity of composted hardwood wastes, it has increased the phytotoxicity of softwood wastes (Worrall 1978).

2.3.2 Sewage and municipal wastes

To assist with the disposal of large quantities of sewage and municipal wastes, city and local authorities frequently make it available to farmers and growers at only a nominal cost. These materials have been successfully used in agriculture and composted sludge has been proved useful in potting mixes, e.g. Chaney *et al.* (1980), Link *et al.* (1983); the usual recommended rate is not more than 30% by volume of the mix. The USDA aerated-pile method of composting transforms sludge into compost in about three weeks, odours are abated and pathogenic organisms are destroyed.

However, composted sludges frequently contain large amounts of heavy metals, the amount often being related to the degree of industrial effluent contamination, e.g. chromium from electro-plating and zinc from galvanizing works.

Nickel, zinc and chromium are often present in high concentrations in sludges and are known to be toxic to plants. Patterson (1971) surveyed the metal content of sewage sludge from 42 areas: the amounts of metals found by extraction in 0.5 M acetic acid are given in Table 2.9.

Nickel is the most toxic of the heavy metals. Toxicities have been reported in soils having above 20 p.p.m. of extractable nickel (soils in which growth is normal have an extractable nickel level of about 5 p.p.m.). Zinc toxicity is probably the most frequently occurring

Table 2.9 Total and 'extractable' amounts of nickel, zinc, chromium and copper found in sewage sludges.

	Total (p.p.m.)	Extractable (p.p.m.)	Extractable (%)
nickel	20–5000	6.8–320	14.7–92.7
zinc	800–49 200	230–7100	14.5–97.4
chromium	40–8800	0.9–170	0.7–8.5
copper	200–8000	2.9–460	0.5–30.9

type following the use of sewage sludge. Comparable levels of extractable zinc (p.p.m.) found in soils showing poor and good plant growth were:

	Poor growth	Good growth
Chrysanthemum	93	37
wallflower	154	56
bedding plants	150	100

Levels at which chromium becomes toxic to plants vary with the form of the chromium and the soil pH. It has been suggested that toxic effects can occur in soil when there is 10 p.p.m. of chromium present as potassium dichromate, whereas with chromium trichloride 100 p.p.m. of chromium are necessary to cause toxicity. Unlike most other metals, chromium is more toxic at high pH levels. Copper is not likely to cause toxicity below 150 p.p.m. of copper in the potting medium. Boron can also be at toxic levels in some municipal wastes. Cadmium can accumulate in sludge-grown plants without being phytotoxic; however, its entry into the food chain is of major concern to human health.

Sterrett *et al*. (1982) concluded that while sludges of low metal content were acceptable, those with high contents presented a severe risk to plant growth. They recommended limits of:

	Zinc	Copper	Nickel	Lead	Cadmium
concentration (p.p.m.)	2500	1000	200	1000	25

Belgian standards for toxic levels of metals in substrates are:

	Concentrations (p.p.m.)								
	Zn	Cu	Ni	Pb	Cd	Mn	Cr	Co	Hg
food crops	300	50	50	300	5	400	25	50	5
ornamentals	500	500	100	500	5	500	200	50	5

There is also a danger in municipal wastes from glass fragments. The deciding factor in the use of sludges in potting mixes will be that of economic savings balanced against the risk of crop loss. In the UK in 1984, the total cost of a peat-based potting mix for growing a pot chrysanthemum was less than 3% of the wholesale value of the plant. Taking risks with substrate toxicities cannot therefore be justified when growing high-value crops.

2.3.3 Animal fibre

Intensive animal management has lead to problems of waste disposal. Slurries often contain too much solid material to be pumped, but this material can be removed by a separator and dried. The dried material composts spontaneously and resembles moss peat. The fibre has a relatively high pH (c. 7.0) and the salt content can be as high as that of a fertilized peat mix. The potassium levels are very high and the levels of nitrogen and phosphorus are also high; pig fibre may also have high levels of copper and zinc. The fibre is not as stable as peat and shows 'shrinkage' in the container. However, when mixed with peat and bark it has given promising results.

2.3.4   Spent mushroom compost (SMC)

This material is characterized by its high pH, usually 7.0 or over, high salinity and high levels of phosphorus and potassium; the nitrogen content is usually between 1.0 and 1.5%. In the fresh state SMC resembles partly-decomposed straw and does not retain much water. Once it is in the container, however, further decomposition and shrinkage occurs quite rapidly, this increases the water retention but reduces the air-filled porosity. Usually, SMC is aged for 9–12 months before being used in potting mixes, at 25–50% of the volume of the mix. The low air-filled porosity of the SMC can be improved by adding a coarse aggregate, e.g. pine bark, perlite, etc. The high pH is countered by adding sulphur; however, it is well buffered against pH change (§ 4.7.2). Iron sulphate applied weekly at a rate of $0.5 \, g \, l^{-1}$ (0.5 lb per 100 gal) will also help to keep the pH below 7.0. Because of the naturally high levels of calcium and trace elements these will not be required in the mix.

2.3.5   Cofuna

Cofuna is a proprietary product produced in France and used in north-west Europe. It is made from a mixture of oil cakes and seaweed which is ground, fermented and inoculated with a mixture of bacteria. These include the cellulolytic, ligninolytic and pectinolytic bacteria which break down fibrous material, and also the nitrogen-fixing bacteria which produce nitrates from the atmospheric nitrogen which plants are otherwise unable to utilize. Cofuna is used in some countries as an alternative to farmyard manure. Its chemical analysis is:

nitrogen (organic)	2.02%	iron	0.14%
nitrogen (ammoniacal)	0.15%	manganese	0.002%
phosphorus	0.23%	copper	0.0003%
potassium	0.51%	zinc	0.01%
sodium	0.13%	chloride	0.10%
calcium	1.94%	molybdenum	9 p.p.m.
magnesium	0.13%	C : N ratio	22.7
pH	7.20		

number of micro-organisms g^{-1} 2 700 000 000

Nelson (1972) compared the growth of chrysanthemums in mixes containing Cofuna with those in other mixes having various soil amendments. Tissue analysis of the leaves of plants grown in a mix having 33% by volume of Cofuna, but which did not receive any liquid fertilizers, showed that the levels of phosphorus, calcium, magnesium and iron were similar to those of plants grown in a mix without Cofuna, but which had been given weekly applications of liquid fertilizers. The nitrogen and potassium levels of the plants grown in the Cofuna mix were between 50 and 70% of those levels in plants which had been given regular liquid feeding. This gives some indication of the amount of plant nutrients supplied by Cofuna. Cofuna should not be steamed – if necessary, other ingredients should be pasteurized separately.

2.4 MINERALS

2.4.1 Sand and gravel

Sand and gravel are seldom used by themselves as the basis of a mix; they are usually used in an admixture with peat, for the purpose of changing the general physical properties, e.g. the bulk density and water retention. Provided that they are free from clay and calcium carbonate, sand and gravel have no effect upon the chemical characteristics of the medium, other than as dilutants. The difference between sand and gravel is purely one of particle size.

In horticultural practice, the term 'fine sand' is colloquially used for sands of particle size between 0.05 and 0.5 mm diameter; this approximates to the 'fine' and 'medium' grades as used in mineral soil particle-size analysis (§ 11.2.1). A typical grade of fine sand used in a peat–sand mix would have a weight analysis of 100% passing through a 40-mesh BS sieve and 40% passing through a 60-mesh sieve.

Although the sand or gravel by itself may have a good drainage rate and low water retention, it does not follow that it will have the same effect when used in potting mixes, much will depend upon the general physical properties of the other constituents. The effects on the air and water relations of varying grades of minerals, and their proportions in peat mixes, are described in the following chapter.

An important point to observe in selecting sand is its freedom from carbonates. These will cause a large rise in the pH of the medium and thereby create nutritional disorders, primarily affecting the availability of minor elements, especially boron and iron. A simple test for the presence of carbonates is to add some dilute hydrochloric acid to the sand; the presence of carbonates is indicated by the frothing and bubbling caused by the production of carbon dioxide.

Sand weighs about 1600 kg m^{-3} (100 lb ft^{-3}) and one of its functions is to increase the bulk density of the mix. This can be an advantage in the case of tall plants and nursery stock, which have a tendency to topple over, especially when grown in lightweight plastic pots. In addition to giving density, the use of the correct grade of sand increases the wettability of the mix.

2.4.2 Clay

Powdered clay is sometimes used in potting mixes, the purpose being to increase the 'buffer capacity' or the resistance to sharp changes in nutrient levels. Clay has a high cation exchange capacity (see Ch. 4), and its prime function is to regulate the supply of phosphorus, exchangeable potassium and minor elements. If the clay is in a powdered rather than an aggregate form it does not improve the physical properties of the mix, indeed there is a reduction in the air-filled pore space if more than a small amount of powdered clay is used. As with sand, it is very important to ensure that it does not contain any carbonates.

2.4.3 Expanded or calcined aggregates

Stable aggregates can be produced when minerals such as clay, pulverized fuel ash, shales, etc., are fired (calcined) at high temperatures. The aggregates have a relatively low bulk density, $0.3–0.7 \text{ g cm}^{-3}$, with an internal porosity of about 40–50%. Their main purpose is to modify the physical properties of potting mixes. Other associated chemical changes in the mix can be a reduction in the amount of water-soluble phosphorus and, with some sources of fuel ash, an increase in the levels of boron and manganese. Use of these materials may require some adjustment to the normal rates of application of plant nutrients. Examples of materials made from clay are: 'Leca' and 'Abmat' in the UK, and 'Gro-sorb', 'Terragreen' and 'Turface' in the USA. 'Haydite' is a combination of clay and shale, while 'Idealite' and 'Nytralite' are shales. 'Hortag' in the UK is made from pulverized fuel ash.

2.4.4 Vermiculite

This is an aluminium–iron–magnesium silicate, which in its natural state is a thin plate-like or laminar mineral resembling mica in appearance. Deposits are found in both the USA and South Africa, and it is more widely used in these countries than in Europe. For horticultural purposes the mineral is first 'exfoliated', a term given to the process of heating previously-graded particles to approximately 1000 °C, usually for about one minute. During this process the water trapped between the layers of mineral is rapidly converted to steam and the resulting increase in pressure causes the plates or layers to expand to 15–20 times their original volume, giving a lattice-like structure. In this form it has a high porosity value and a good air–water relationship.

The material is available in a number of grades, ranging from a fine particle grade for seed germination, up to a grade with particles 6 mm ($\frac{1}{4}$ in) in diameter; the average density is only about 80 kg m^{-3} (5 lb ft^{-3}). Vermiculite has a relatively high cation exchange capacity, about 100–150 meq 100 g^{-1} and compares favourably with peat in this respect. Most samples contain 5–8% of available potassium and 9–12% of magnesium; mixes containing vermiculite therefore require less of these minerals in the base fertilizer.

From the horticultural viewpoint vermiculite can be classified into two types: one is naturally slightly acid in reaction with a pH of about 6.0–6.8; the other type contains a significant amount of magnesium limestone which is broken into small pieces during exfoliation and thus raises the pH to above the neutral point. The former type of mineral is preferable for plant growing because of the usual difficulties with mineral nutrition which occur in alkaline conditions.

Koths & Adzima (1978) compared vermiculites from the USA and South Africa. The African vermiculite was more alkaline than the American material, the pH values being 9.3–9.7 and 6.3–7.8 respectively. The African vermiculite also had much higher magnesium levels: the Ca:Mg ratio as measured by a Morgan's extract was 1:2 compared with a 6:1 ratio for the American product. It was concluded that the African vermiculite was not detrimental when mixed with peat, but if lime was required a calcitic limestone rather than a Dolomite limestone should be used. Nelson (1969) examined ways of reducing the pH of Palabora vermiculite. If the vermiculite is not to be mixed with a low pH sphagnum peat, a phosphoric acid drench at 0.68 ml of 85% phosphoric acid per 100 g of dry Palabora vermiculite (no. 3 grade) was recommended. However, this treatment should not be applied if slow-release fertilizers have been incorporated.

Vermiculite does not adsorb the anions Cl^-, NO_3^- and SO_4^{2-} but it does adsorb some PO_4^{3-}. Bylov *et al.* (1971) have reported that when vermiculite was treated with a solution of potassium dihydrogen phosphate, between 63 and 77% of the phosphorus was adsorbed, the actual amount depending upon the type of vermiculite. Only 25% of the adsorbed phosphorus was retained in an available form and 75% was in a fixed or unavoidable form. The phosphorus formed insoluble compounds with sesquioxides and magnesium. Vermiculite is also able to 'fix' large quantities of ammonium in an unavailable form; this is useful in regulating the amount of nitrogen available to plants when large amounts of organic or ammonium-producing fertilizers are used. Most of the 'fixed' ammonium is, however, available to bacteria and is converted to nitrate within a few weeks. It is then available to plants.

When vermiculite is used by itself as a growing medium for long-term cropping, there is a tendency for the lattices or honeycomb structure to collapse, resulting in reduced aeration and drainage and thus rendering the material 'soggy'. For this reason it is advisable to mix either some perlite or peat with the vermiculite as is done with the Cornell 'peat–lite' mixes (Boodley & Sheldrake 1972).

2.4.5 Perlite

Perlite is an alumino silicate of volcanic origin and is widely used in the USA and New Zealand, both countries having large natural deposits of this mineral. When crushed and heated rapidly to 1000 °C it expands to form white, lightweight aggregates with a closed cellular structure; these aggregates are stable and do not break down in the mix. The average density of perlite is 128 kg m^{-3} (8 lb ft^{-3}) and it is available in a range of graded particle sizes. Because of its closed cellular structure, water is retained only on the surface of the aggregates or in the pore spaces between the aggregates. This means that mixes with a high proportion of perlite are usually well drained and do not retain much water. Warren Wilson & Tunny (1965) found that the volume of water present between container capacity and the wilting point was 7% for perlite and 42% for vermiculite. When plants were grown in perlite, a capillary watering system gave better results than frequent hand watering. White & Mastalerz (1966) found water availability values of 34% for perlite and 59% for peat; the available water content of perlite–peat mixtures increased as the perlite content was decreased and the peat correspondingly increased. The actual amount of water retained by any medium is of course largely dependent upon the pore size distribution and the actual water availability in perlite will depend upon the grade of material used. It is general experience, however, that mixes containing perlite are well aerated and have a lower available water content. For this reason perlite is often mixed with peat and used as a rooting medium for cuttings, because its open structure prevents the occurrence of waterlogging from the frequent mist application of water to the cuttings during rooting. The low bulk density of the mixture also means that the delicate roots are not so easily broken off during handling.

The chemical characteristics of perlite can be summarized by saying that it has virtually no cation exchange capacity; Morrison et al. (1960) reported a value of only 1.5 meq 100 g^{-1}. It is composed mostly of silicon dioxide (73%) and aluminium oxide (13%), and for practical purposes it can be considered devoid of plant nutrients. Plants grown in mixes containing large amounts of perlite are, therefore, largely dependent upon liquid feeding. Green (1968) found that carnations grown in perlite suffered from aluminium toxicity when the pH of the nutrient solution was below 5.0; above this value no toxicity was observed. Perlite contains small amounts of fluoride, and certain plants are susceptible to fluoride injury. However, the amount of fluoride in perlite (17 p.p.m.) is very low in relation to the levels in some other materials used in potting mixes, e.g. Sheldrake et al. (1978) found 1254 p.p.m. fluoride in superphosphate. The level of soluble fluorides can be reduced by liming, and it is unlikely that plants growing in limed peat–perlite mixes will show any effects of fluoride from the perlite. If fluoride toxicity does arise then the source of phosphorus should be checked (§ 6.1.5).

2.4.6 Mineral or rock wool

Slabs of mineral or rock wool are now extensively used in Europe for tomato and cucumber production. It is also available in blocks or cubes for rooting cuttings and growing young plants. This material, which is largely an alumino silicate with some calcium and magnesium also present, is smelted to approximately 1500 °C and fibres are formed from the molten material as it is cooled. In its prepared state it has a pore volume of about 97% and its function is to provide root anchorage for the plant and to regulate the water and air supply. It does not contain any plant nutrients and the plant must rely entirely on the inclusion of nutrients in the water supply. This system of growing is in fact a modification of the sand or gravel culture system.

Mineral wool in granules or flocks, 3–6 mm diameter, is also used to improve the structure of potting mixes. The wool is available in both water absorbent and water repellent forms. In the latter form, air is retained within the granule even when the surrounding media is water saturated, thereby providing some aeration for roots present within the granule.

2.4.7 Pumice

Pumice is an alumino silicate of volcanic origin, it contains some potassium and sodium but only traces of calcium, magnesium and iron. The material is porous, the pores being formed by the escape of steam or gas when the lava is cooling. It is sometimes used as a physical conditioner in potting mixes or as an alternative to sand or gravel in hydroponic cultures. However, the particles are not very stable and break down easily. In its natural state the material contains only small amounts of plant nutrients. It is, however, able to absorb some calcium, potassium, magnesium and phosphorus from the soil solution, and release them to the plant later.

2.4.8 Zeolites

The zeolite group of minerals are alumino silicates formed from weathered volcanic rocks. They have a honeycomb structure with very small pores (0.5 nm) that allows K^+ and NH_4^+ ions to enter the granule, but excludes the bacteria which convert ammonium salts to nitrates. Zeolites have a high cation exchange capacity and act as slow-release ammonium and potassium fertilizers. A mixture of 90% sphagnum peat and 10% zeolite has an exchange capacity of 290 meq l^{-1} or double that of peat alone. Hershey et al. (1980) found that 50 g of the zeolite, clinoptilolite, per 1.5 l of potting mix produced pot chrysanthemums equal to those irrigated daily with a solution containing 234 p.p.m. of potassium. Unlike clays, clinoptilolite does not absorb phosphorus.

2.5   PLASTICS

One result of the rapid growth of the plastics industry has been the increasing interest shown in the use of foam plastics in pot plant mixes. Initially waste polystyrene from industrial processes was used to make plastic flower pots and within a few years this type of pot had virtually displaced the traditional porous clay pot (Ch. 14). More recently, growers have been experimenting with mixes made from peat and plastic materials in foam or expanded form.

2.5.1 Expanded polystyrene flakes

Flakes of expanded polystyrene 4–12 mm in diameter were first used in agriculture for draining and improving the physical condition of heavy clay soils. The material is chemically neutral, it does not decompose or compress in normal use and has a low density of only 20 kg m^{-3} (34 lb yd^{-3}). Flakes are formed from a large number of small closed cells and although the total porosity of the material may be as high as 95% water is not absorbed within the flake. Thus, when used in potting mixes it has the effect of reducing the amount of water retained, and because of the particle sizes, improving the aeration. Usually, the polystyrene is mixed with peat, the actual ratio of polystyrene to peat being adjusted to the type of plant grown. For example, mixes for *Cyclamen*, *Gloxinia*, *Fuchsia*, etc., usually contain 25% by volume of polystyrene and 75% of peat, whereas for epiphytic plants which require less water, the polystyrene content is increased to 50%.

It is important to remember that this material does not contain any plant nutrients and neither will it absorb nor retain any from the fertilizers. Some allowance must therefore be made by starting liquid feeding earlier than usual. Mixes made with this material must not be steamed and neither should chemical sterilizers such as chloropicrin and methyl bromide be used. Minor disadvantages of this material are its electrostatic properties during mixing and the tendency for large particles to rise to the surface of the mix following irrigation.

2.5.2 Urea-formaldehyde foam resins

From the horticultural viewpoint the principal difference between expanded polystyrene and urea-formaldehyde foams is the ability of the latter material to absorb water. It has an open cellular structure and can absorb between 50 and 70% of its volume of water. It has a low density, 10–30 kg m^{-3} (16–50 lb yd^{-3}), and is available in both flake and block form. The formaldehyde content is less than 2.5% but it is important to ensure that any free formaldehyde which may be present in the recently manufactured material is allowed to evaporate before the foam is used in making potting mixes.

Urea-formaldehyde foams do not have such high stability in soils as expanded polystyrene. Under acid conditions the annual rate of degradation is 15–20% and the 30% by weight of nitrogen in the foam then becomes available to the plants during decomposition; there is also a very small amount of free nitrogen (0.25%) present in the material after manufacture. However, because of the foam's very low density and slow rate of decomposition, the rate at which nitrogen is made available to the plants is not very high. The material has negligible amounts of other nutrients and the pH is about 3.0. Thus although the physical properties of urea-formaldehyde are somewhat similar to peat, the chemical or nutritional characteristics of the two materials are quite dissimilar. It is normally used in mixes at 20–50% by volume and, as with the expanded polystyrene, careful attention to plant nutrition is required.

This material is sold in Germany under the trade name 'Hygromull' and as 'Floramull' in the USA.

2.5.3 Polyurethane foam

Another group of foams are made from polyurethane. In common with other foams, they have a low bulk density, i.e. 12–15 kg m^{-3} (20–25 lb yd^{-3}). The open cellular structure gives a maximum water-retaining capacity of 70% of its volume. The pH of the foam is approximately neutral, and it is not decomposed by micro-organisms, nor does it contain any plant nutrients. It is available in the usual flake form and also as cubes or blocks with a shallow depression in the upper surface into which seedlings or cuttings can be inserted. The blocks or cubes are usually partially joined together at the base to form a sheet and are easily separated when the plants require spacing. The blocks readily absorb water, either by capillary absorption or by overhead spraying. As the material does not contain any nutrients, all the plants' requirements must be supplied by liquid feeding.

To ensure freedom from aromatic amines, which can cause phytotox-icity, Wheeler *et al.* (1985) recommended either rinsing polyurethane foam plugs in ethanol and then water, or heating the plugs to 100 °C for 2 hours. A polyurethane form in block form – 'Plant-In', made by Smithers-Oasis – is available in western Europe.

2.5.4 Phenolic resin foam

A recent addition to the range of foam substrates is a phenolic resin foam. The bulk of this material (28 kg m^{-3}) is greater than that of polyurethane foam, but these substances are otherwise similar.

The effects of the various foam plastics in pot-plant mixes on the air–water relationships have been reported by De Boodt & Verdonck (1971).

2.5.5 Ion exchange resins

Water-soluble fertilizers create two types of problem in loamless mixes. They can result in high salinity or osmotic potentials and they can be easily removed by leaching from many types of media (§ 4.3.2). These problems have been approached experimentally by the use of mixed cation and anion exchange resins. The cation exchange resins retain K$^+$, NH$_4^+$, Ca^{2+}, etc. and the anion exchange resins NO$_3^-$, PO$_4^{3-}$, etc. in an exchangeable form.

Nutrient-impregnated resins have been used in mixes at rates of 2–10% by volume, the rate being adjusted to the anticipated demand for nut-rients by the plant and the length of the growing period. Nutrients are released from the resin by an exchange with other ions already present in the irrigation water. To be effective, the irrigation water must have sufficient salts present to give a salinity of between 200 and 1200 μmho

Table 2.10 Dry weights and porosity of some potting mix materials.

Material	Dry weight ($kg\,m^{-3}$)	($lb\,ft^{-3}$)	Porosity (%)
bark, fir (<3 mm)	224	14	80
bark, redwood (<10 mm)	128	8	85
mineral wool	65	4	98
peat, sphagnum	104	6.5	93
peat, sedge	224	14	85
perlite	96–128	6–8	75
plastic foams	8–32	0.5–2	variable
pumice	480	30	65
rice hulls	104	6.5	80
sand	1600	100	40
sawdust	192	12	78
vermiculite	80–112	5–7	80

(0.2–1.2 m mho); if the salinity falls below 200 μmho, the rate of ion release from the resin may not be sufficient to meet the plant's requirements. At present there is insufficient experimental data upon which to base recommendations for the general use of nutrient-impregnated resins in pot-plant mixes. However, ion exchange resins, such as 'Lewatit HD5', are extensively used for the hydroculture of ornamentals (Aguila & Martinez 1980).

Typical dry weight values of some newer materials used in mixes are given in Table 2.10.

CHAPTER THREE

Physical aspects

From the physical viewpoint, plant substrates can be regarded as comprising solid matter interspaced with voids or pores. The substrate has two main physical functions, i.e. to provide the anchorage that enables the plant to support itself, and to regulate the supply of water and oxygen to the roots. The growing of plants in small, shallow containers creates two physical problems. First, the volume of substrate and water available to each plant is small by comparison with the volume available to field crops. Secondly, because the substrate is in a shallow layer, with the base of the container breaking the continuity present in a normal soil profile, a 'perched' water table is created (Spomer 1975a). This prevents the normal amount of drainage and makes the media wetter after irrigation than it would otherwise be, thus rendering the plant liable to inadequate aeration after every irrigation. The other extreme condition of a shortage of water can also occur very quickly under conditions of high evapotranspiration.

It has long been recognized that normal field soils are not suitable for container media unless their structure is modified by adding bulky physical conditioners (Lindley 1855). The desirable physical characteristics of container media can best be described in terms of their bulk density, total pore space, water retention and air-filled porosity.

3.1 PHYSICAL TERMINOLOGY

Before examining the medium from the physical viewpoint it is necessary to define some of the more important physical terms used:

air-filled porosity (AFP) (air capacity)   The proportion of the volume of medium that contains air after it has been saturated with water and allowed to drain.

available water holding capacity (AWHC)   The amount of water present after the medium has been saturated and allowed to drain, less the amount still present at the permanent wilting percentage.

bulk density   The dry mass per unit volume of moist medium.

capillary water   Water that is retained in the small pores of the medium by surface tension and moves as a result of capillary forces.

container capacity   The total amount of water present after the medium in a container has been saturated and allowed to drain. The amount of water retained will be greater than that present in an agricultural soil at 'field capacity'.

desorption curve   See soil water tension curve.

easily available water (EAW)   The amount of water held by a medium after it has been saturated with water and allowed to drain, minus the amount of water present at some defined water tension. This is often taken as the volume of water between 10 and 50 cm tensions.

matric potential (ψ_m)   The energy or negative pressure with which water is retained in the soil. It is the force per unit area that must be applied to remove water from the soil at any given water content. It does not include the osmotic potential or salt effect.

osmotic potential (ψ_s)   The decrease in the free energy of the soil water caused by the presence of dissolved salts. It is the negative pressure to which water must be subjected in order to be at equilibrium with the soil solution through a semipermeable membrane.

oxygen diffusion rate (ODR)   The rate at which oxygen diffuses through media to the roots of plants. It is usually measured with a platinum micro-electrode.

particle density (specific gravity)   The ratio of the bulk density or volume weight of the medium to the mass of unit volume of water.

perched water table   The saturated layer of medium in the base of a container after drainage ceases.

permanent wilting percentage   The amount of water still present in the medium when wilting plants are unable to extract sufficient water to regain turgidity even when placed in an atmosphere saturated with water vapour.

pF   The logarithm of the soil water tension expressed in centimetres height of a column of water, e.g. 100 cm water tension = pF 2 and the permanent wilting percentage = 16 000 cm water tension = pF 4.2.

soil water tension   See matric potential.

soil water tension curve   A graph showing how the amount of water retained by a soil varies with tension or applied suction pressure. The units used are those of pressure; the low tensions present in wet soils are usually measured in terms of the equivalent heights of columns of water or mercury, and the higher tensions in drier soils are expressed in atmospheres, bars or kPa.

solute stress   See osmotic tension.

total moisture stress (ψ_T)   The sum of the soil matric potential or water tension and the osmotic potential or suction. In soils with very low nutrient levels there will be little difference between the soil water tension and the total soil moisture stress. In horticultural media with high nutrient levels, and especially with loamless mixes, the osmotic (or solute) potential will often exceed the soil water potential (or matric tension) and thus form the most important part of the total soil moisture stress.

total pore space   The total volume of medium not occupied by mineral or organic particles.

volume weight See bulk density.

In addition to their obvious purpose of providing the basis for a physical description of media, some of these parameters are also important for chemical or nutritional reasons. For example, the bulk density of a medium is an important factor to consider in interpreting the results of a chemical analysis if this has been made on the traditional weight basis and the results expressed as parts per million by weight, rather than on a volume basis (§ 4.4).

3.2 BULK DENSITY AND TOTAL PORE SPACE

Bulk density (BD) and total pore space (TPS) are inversely related, and it is convenient to consider them together. Mineral soils and loam-based potting media have high BDs, e.g. the BD of loam-based John Innes compost is about $1.0\,\mathrm{g\,cm^{-3}}$, whereas those of peat and lightweight mixes are about $0.1\,\mathrm{g\,cm^{-3}}$. Mixes having a low BD have not generally been found to be disadvantageous; on the contrary, their lower weight is regarded as an advantage during mixing and when transporting plants. However, a low BD can cause instability when tall plants, e.g. standard fuchsia, are grown in light plastic containers, or when hardy nursery stock are grown on an unsheltered site. When more stability is required the BD of the mix can be increased by the addition of sand or grit. It can also be increased by compaction during potting; however, this has several other effects, some of which are detrimental to plant growth (§ 3.7).

The TPS of substrates ranges from less than 50% of the volume for some mineral soils to about 95% for some peats. The TPS of a medium is calculated from:

$$\text{TPS } (\%) = \left(1 - \frac{\text{bulk density}}{\text{particle density}}\right) \times 100$$

Several workers have shown an empirical relationship between bulk density and total porosity (Beardsell et al. 1979, Hanan et al. 1981, Bunt 1984). From 32 mixes made with peat and minerals in various combinations, having BDs ranging from $0.09\text{--}1.50\,\mathrm{g\,cm^{-3}}$, Bunt (1984) obtained the following relationship (Fig. 3.1):

$$\text{TPS} = 98.39(\pm 0.26) - 36.55(\pm 0.36) \times \text{BD}$$

Mixes containing perlite also fitted this relationship. However, most of the particles of perlite have closed pores so it is necessary to distinguish here between the *total* and *effective* pore space, the latter being the volume available for the exchange of gas and water. The difference between these two categories of pore space in mixes made with varying amounts of a coarse grade of perlite is shown in Figure 3.2. Although the effective pore space was lower than the total pore space, mixes of peat and perlite nevertheless have high porosities. A mix having a high porosity has the potential advantages of good water retention and aeration. Whether this is achieved in practice will, however, depend upon the size distribution of the pores.

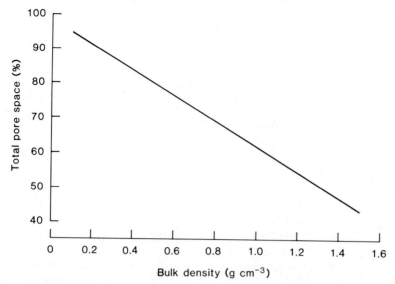

Figure 3.1 Relationship between the total pore space and the bulk density of 32 mixes (r = −0.99).

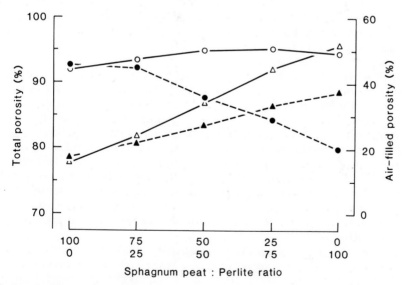

Figure 3.2 Effect of perlite on the *total* and *effective* air-filled porosity and pore space when used in varying ratios with peat.
Key: ○ total pore space; ● effective pore space; △ air-filled porosity; ▲ effective air-filled porosity.

3.3   WATER RELATIONS

3.3.1   Available water holding capacity (AWHC)

Because the volume of medium in which container plants are grown is usually small in relation to the potential loss of water by evapotranspiration, it is desirable that the medium should have a large available water holding capacity. This is defined as the amount of water held between container capacity, i.e. after drainage ceases, and the permanent wilting point, i.e. when the plant is unable to extract any more water from the media. Beardsell *et al.* (1979b) determined the AWHC of a number of materials:

	Peat moss	Poppy straw	Sawdust	Scoria	Pine bark	Sand
AWHC	55.4	37.4	32.2	27.3	25.9	16.7
(% by volume)						

A medium may have a low AWHC because: (a) its total porosity is low, (b) the pores are large and much of the water is lost by gravity, (c) the pores are very small and the plant is unable to extract much of the water before it wilts, or (d) a combination of these factors. The availability of water in seven species of hardwood bark was determined by Spomer (1975b). He found that when saturated the average water content of the barks was about 40% of their wet volume, but that only 25% of this water was available to plants because movement of water from the inner small pores to the outer regions of the particles was very slow.

In another study, Beardsell *et al.* (1979b) found that the time taken by plants to wilt was not necessarily proportional to the water content of the media as determined by physical measurements. Although peat moss has a high AWHC, plants growing in this medium wilted in 10.4 days, whereas plants growing in poppy straw (having a lower AWHC) wilted in 12 days. In the early part of the experiment before the containers dried out, transpiration losses by plants were highest for those growing in peat, intermediate for those in pine bark and lowest for those in poppy straw. A large part of the available water in peat is held at relatively low tensions, and transpiration is maintained at high rates until much of this water is exhausted. In media with small pores the water is less available to the plants, and the rate of transpiration per unit of leaf area will be reduced sooner. Most experimental work on water loss from container plants shows a progressive decline in the rate of transpiration as the soil water deficit increases. Closure by the plant of the leaf stomates to reduce transpirational loss however also reduces the rates of cell growth and photosynthesis.

The growth rates of plants in small containers of fertile media are significantly reduced long before the water content approaches the wilting point. For example, Spomer & Langhans (1975) found that there was a general increase in the growth of chrysanthemums as the water content of the medium was increased to about 90% of pore saturation. The salinity level of potting mixes is high in relation to field soils, this also reduces the availability of water to the plants and thereby reduces their growth rates.

3.3.2 Energy concept of water in media

The availability of water to plants is often considered in terms of the equivalent moisture stress, i.e. the energy required to remove water from the media. The total soil moisture stress (TSMS) has two components, the matric, or water, potential and the solute, or osmotic, potential:

$$\text{TSMS} = \text{matric tension} + \text{osmotic stress}$$
$$\psi_T = \psi_m + \psi_s$$

Both of these components can have large effects on plants grown in containers. (Osmotic stress is discussed in § 4.8.)

Matric potential   The range of matric potentials of greatest interest to pot plant growers (below one atmosphere) can be measured in the laboratory by a sintered plate apparatus or in containers by tensiometers. The particles in a mix vary considerably in diameter, shape and texture, consequently the pores between the particles also vary in diameter and shape. The relationship between the matric potential, or force required to remove the water, and the diameter of the pores is shown in Figure 3.3. As the pore diameter falls below 100 μm the energy required increases sharply. The matric potential range in which pot plants are normally grown, i.e. 10–100 mbars, corresponds to pore diameters in the range

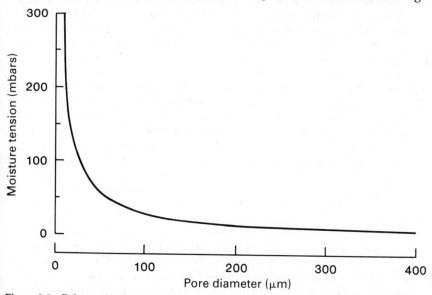

Figure 3.3   Relationship between the matric potential, or force necessary to remove water, and the diameter of pores in a medium.

300–30 μm. After irrigation the large pores lose some of their water by gravitational force, thereby providing aeration for the roots. As more water is removed from the medium it is withdrawn from pores of progressively smaller size. However, the pores are not drained completely, a film of water is retained around the particles. The film decreases in thickness as the medium dries out.

Units of measurement Matric potential can be expressed in various units of negative pressure. In much of the early work on soil water relations the basic unit of measurement was cm of water tension. As agricultural crops are grown with a much wider range of soil moisture contents than crops in containers, this can result in working with very high values of moisture tension. To avoid this, the common logarithm of the tension is used, denoted by the symbol pF. In this way the moisture content at the permanent wilting point is expressed as pF 4.2 rather than 16 000 cm. For container media, desorption curves are often presented over the range 0–100 cm of water tension. Sometimes the units of tension are quoted in bars or atmospheres; for *practical purposes* the bar can be regarded as equivalent to one atmosphere or 1000 cm of water, errors resulting from this approximation will be small. Tensions can also be expressed as cm of mercury or in the SI unit the pascal (Pa), 1 kPa is approximately 10 cm water tension; relationships between these units are given in Table 3.1.

Although weight can be measured more easily and accurately than volume, the wide variation in bulk densities of potting mixes means that useful comparisons of water relations in different media can only be made on a volume basis. An example of a desorption curve over the range pF 0–7.0 for a mix of 90% peat moss and 10% sand is shown in Figure 3.4. At pF 1.0 (10 cm water tension) this medium had about 10% of air-filled porosity, and at the permanent wilting point it still contained about 16% by volume of water.

Container depth and water retention The depth of the container has a considerable effect on the water:air contents of a mix. After a container of potting medium has been irrigated and allowed to drain, the water potential at the base will be at zero; for each centimetre in height of

Table 3.1 Approximate relationships between units used to measure soil water tension.

cm of water	pF	atmosphere	Unit mbars	KPa	cm of mercury
1	0	0.000966	0.9789	0.098	0.0734
10	1	0.00966	9.789	0.978	0.734
100	2	0.0966	97.89	9.78	7.34
1000	3	0.966	978.9	97.8	73.4
15000	4.17	14.49	14683	1467	1101

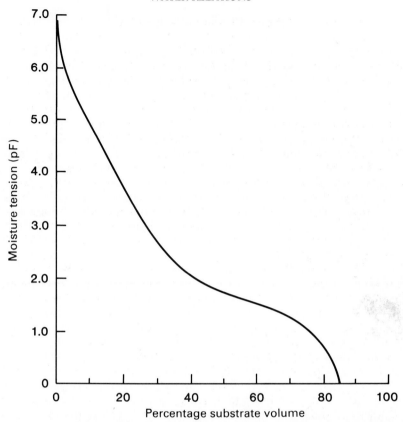

Figure 3.4 Desorption or water release curve of a potting mix made from 90% peat moss 10% sand (after Arnold Bik 1973). Most of the water was available at tensions below pF 2.0 (100 cm of water) but about 16% of the water was still present at the permanent wilting point (pF 4.2).

the medium in the container the tension will increase by 1 cm of water. In a container with 10 cm depth of medium the tension would therefore range from 0–10 cm of water, with the average tension being 5 cm of water. The effects which this range of tensions will have on the volumes of air and water at different depths in the medium will depend upon the pore size distribution. For example, Figures 3.9 a & b show that virtually no water was released from the fine sand at tensions below 20 cm of water, whereas in the coarse grit almost all the water had been released at 5 cm of water. However, these are extreme examples, with most growing media there will be a progressive decrease in water retention as the height of the medium is increased. In practical terms it means that potting mixes in shallow layers, e.g. in seed boxes or flats, will be wetter than those in deeper containers.

3.4 AERATION OF SUBSTRATES

Plant roots require an adequate supply of air as well as water. Under conditions of low evapotranspiration, as in winter, lack of aeration for the roots of container plants can be as great a problem as a shortage of water in summer. Roots require oxygen to maintain their metabolic activity and growth; a temporary shortage can reduce root and shoot growth, and anaerobic conditions for only a few days will result in the death of some roots. It is estimated that under field conditions roots can consume up to nine times their volume of gaseous oxygen each day. Oxygen is also required by soil micro-organisms, and plants growing in peat with a high microbial population can require twice as much oxygen as plants in a sandy loam. The temperature of the medium also affects the oxygen status in two ways: an increase of 10 °C in the temperature doubles the respiration rate; at the same time an increase in the temperature of the soil water from 20–30 °C reduces the amount of oxygen that can be dissolved from $9.6 \, mg \, l^{-1}$ to $7.8 \, mg \, l^{-1}$.

Plant roots are normally covered with a film of water, and although some of the oxygen they require will already be dissolved in the irrigation water, this will only be a fraction of the total requirement. Most of the oxygen needed by the roots must therefore diffuse down through the air-filled channels in the substrate and then across the water film to the roots. Usually it is the diffusion of oxygen across the water film that is the limiting factor to root aeration. At 20 °C, the diffusion coefficient for oxygen in air is $0.214 \, cm^{-2} \, s^{-1}$, and that for oxygen in water is $2.22 \times 10^{-5} \, cm^{-2} \, s^{-1}$; i.e. oxygen diffuses approximately 10 000 times more slowly in water than in air. The thicker the film of water around the roots, the greater the reduction in the concentration of oxygen at the root surface will be. There will be no difficulty in maintaining normal oxygen concentrations in the bulk of the soil atmosphere provided that the pores or channels are continuous to the surface of the medium and are not blocked by water. Localized deficiencies of oxygen can occur, even if the medium contains air-filled channels, as long as some of the aggregates with fine pores remain in a saturated condition. It has been estimated that when respiration rates are high, it is possible to have an oxygen concentration of zero inside soil aggregates of 0.1–1.0 cm radius, although the water surrounding the aggregates is fully saturated with air (Greenwood 1969).

Another way in which some plants obtain their oxygen is by conducting it from the leaves down through the stems and into the roots. This method of supply is well established for some cereals; however, there is less information on horticultural crops.

3.4.1 Air-filled porosity (AFP)

The AFP of potting mixes is often quoted as an index of aeration. Although there is no precise agreement on an optimal value, or even on a narrow range of values, there is a general consensus that 10–20% of AFP is

Table 3.2 Approximate root aeration requirements of selected ornamentals, expressed as the free porosity (Johnson 1968).

aeration requirements:	very high	high	intermediate	low
free porosity (%):	20	20–10	10–5	5–2
	Azalea	Antirrhinum	Camellia	carnation
	orchid	Begonia	Chrysanthemum	conifer
	(epiphytic)	Daphne	Gladiolus	geranium
		Erica	Hydrangea	ivy
		foliage plants	lily	palm
		Gardinia	Poinsettia	rose
		Gloxinia		stock
		orchid		Strelitzia
		(terrestrial)		turf
		Podocarpus		
		Rhododendron		
		Saintpaulia		

desirable. This lack of precise agreement between researchers on an optimal AFP for container media can be attributed to:

(a) differing tolerances of plant species to low levels of soil aeration;
(b) the effects of different management and environmental factors;
(c) the different methods used for determination of AFP

Some plants require good aeration if root damage is to be avoided, e.g. *Brassaia actinophylla*, *Garrya elliptica*, *Rhododendron* spp. and *Erica* spp.; others are more tolerant of less aerated media, e.g. carnation, *Cyperus alternifolius* and *Cornus* spp. Further examples of the different AFP requirements of plants are given in Table 3.2.

The adverse effects of growing plants in media with poor aeration can be at least partially mitigated by good management and environment. If frequent and heavy irrigations are avoided the AFP will be enhanced, also under high rates of evapotranspiration the rapid loss of water from the medium soon creates an adequate AFP. Dasberg & Bakker (1970) used several indices to relate plant growth with air content of the media. They obtained the best correlation using the *mean daily* or *time integrated* air content as an index of aeration. Although high AFP values are not essential for a potting mix, such values make management much easier and decrease the risk of overwatering. An example of plant response under poor management to a range of AFP values is shown in Figure 3.5. Tomato plants were grown in winter in two sets of similar substrates; one set was irrigated each day irrespective of the need

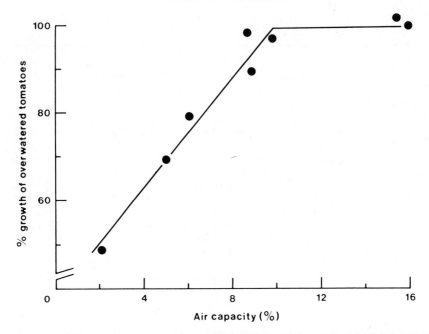

Figure 3.5 Relationship between the air-filled porosity of the medium and the risk of overwatering. Plants grown in mixes with low air-filled porosities and watered too frequently showed reduced growth, in comparison with plants grown in similar mixes and given careful watering. When the air-filled porosity of the mix was 10% or greater, there was no reduction in plant growth at the high frequency of watering.

for water, the other set received careful management. Leachate was returned at the next irrigation to avoid nutritional affects. Growth of the overwatered plants was expressed as a percentage of the growth of plants receiving good management. In this experiment growth declined progressively in overwatered substrates as the AFP fell below 10%.

Two methods commonly used to determine the AFPs of media are:

(a) calculation of AFP from the total porosity and the weight of water present after saturation and drainage;
(b) from desorption curves, AFP is calculated at a given moisture tension.

Each method has advantages and disadvantages, and will not always give similar results. In method (a) the containers are irrigated twice daily for eight weeks; this increases the bulk density of the media and reduces the volume of macro-pores, also fine particles are moved down the profile causing some 'plugging'. Although these conditions are similar to those experienced by plants, it may still not be possible to saturate hydrophobic materials by the normal procedure of wetting under tension and submerging for 12 hours, therefore for a more accurate estimation of AFP a

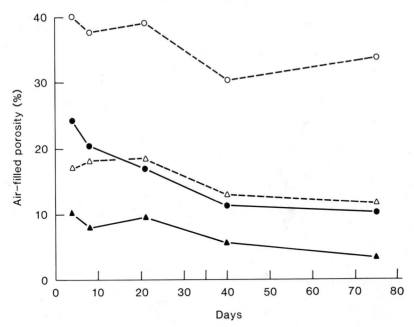

Days

Figure 3.6 Effects of different watering methods and application of surfactant on the air-filled porosity of a hydrophobic peat. Surface watering gave much higher AFP values than did soaking; the surfactant significantly increased wetting and reduced the AFP. Peat, surface watered twice daily without a surfactant treatment, was still resistant to wetting after 75 days.
Key: ○ surface watered to excess, no surfactant; ● surface watered to excess, one application of surfactant; △ wetted under tension and soaked overnight, no surfactant; ▲ wetted under tension and soaked overnight, one application of surfactant.

surfactant may be required (Fig. 3.6). In method (b) the initial bulk density of the medium in the sintered glass funnel will be less than that in containers. Also the shallow layer of medium (c. 1.5 cm) shrinks relatively more than a similar medium in containers. In method (b) the AFP is sometimes calculated at a tension of 10 cm of water (pF 1.0) and sometimes at 32 cm of water (pF 1.5). In the container method the tension will be equivalent to half the height of the container, i.e. it will often be equal to 5 cm of water.

In a third method the AFP is more rapidly estimated from the volume of water released from a known volume of media that has been saturated and allowed to drain (Bragg & Chambers 1987).

3.4.2 Oxygen diffusion rate (ODR)

Another index of the aeration status of potting mixes is the rate of oxygen diffusion. To measure this, a small platinum electrode, which simulates a plant root, is inserted into the medium and an electrical potential applied between it and a calomel reference electrode. The rate at which oxygen is reduced at the electrode surface, after diffusing through the gas-filled channels and across the water film covering the electrode, is a measure of the oxygen-supplying power of the medium. The small size of the platinum electrode, usually 1 cm in length, enables ODR determinations to be made at different positions and depths, thereby creating an ODR profile of the medium in the container.

Some examples of ODR profiles of media made with materials of different particle sizes are shown in Figure 3.7. With each mix the ODR decreased substantially as the depth increased; this was due to the medium at the bottom of the container being wetter – diffusion through the thicker films of water was slower. Therefore, the greatest concentration of roots will normally be in the zone having the lowest ODR. Mixes containing coarse particles had much higher ODR and air-filled porosity values than those made with fine particles. The following relationship was obtained between the mean ODR of 13 media and their mean AFP values (Bunt *et al*. 1987):

$$\text{ODR} = 10 + \text{AFP}^{1.85}$$

From this relationship the ODR at 10% AFP = 80×10^{-8} g O_2 cm^{-2} min^{-1}. Paul & Lee (1976) grew chrysanthemums in 13 different media and correlated plant growth with the mean ODR of each substrate. Over the

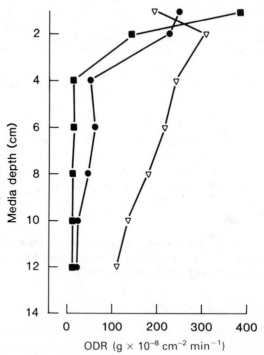

Figure 3.7 Oxygen diffusion profiles of media in 14-cm-deep containers. *Key:* ▽ 75% peat 25% coarse sand; ● 75% peat 25% fine sand; ■ mix made from 1/3 peat 1/3 redwood 1/3 fine sand.

range $5 \times 10^{-8} - 40 \times 10^{-8}$ g O_2 cm^{-2} min^{-1} there was an almost linear response in plant growth, which continued to increase to the highest ODR value of 65×10^{-8} g O_2 cm^{-2} min^{-1}. This value is almost twice as high as the reported optimal value for field crops. However, this is to be expected because container plants grown in greenhouses have much higher root densities, nutrient levels and growth rates than field crops, so their requirements for oxygen will also be much higher.

3.4.3 Chemical reactions in anaerobic media

As well as the direct effects oxygen deficiency has on root and shoot growth, other indirect effects include the production of substances toxic to plants, e.g. ethanol and ethylene, and the reduction of manganese and sulphur compounds. Ethylene can occur at concentrations of up to 10 p.p.m. in waterlogged soils, and for some species root growth is restricted at 2 p.p.m. Under anaerobic conditions sensitive plants, e.g. tomato, may have three times the normal concentration of ethylene in their shoots, causing epinastic curling of the leaves. Although carbon dioxide can accumulate in anaerobic media it is not usually considered to adversely affect plant growth, largely because it is almost 300 times more soluble in water than oxygen and is therefore able to diffuse away from the roots readily, down a concentration gradient. Plants grown in some soil-based media that are kept too wet can suffer from manganese toxicity, as under the anaerobic conditions manganese is reduced to the divalent form. Organic sulphur can also be reduced to sulphides by bacterial action if the medium is anaerobic.

3.5 FORMULATING MIXES

Although there is no universally accepted 'optimal' physical specification for potting mixes, as this must reflect the varying demands of plant species, environment and management practices, there is a general agreement on an acceptable range of physical parameters for greenhouse container media. For example, the recommendations of:

(a) De Boodt & Verdonck (1972):

total porespace	85%	
airspace	20–30%	airspace determined at 10 cm water tension
easily available water	20–30%	easily available water = volume between 10 and 50 cm water tension determinations made without compaction or settling of the media by irrigation
water buffer capacity	4–10%	water buffer capacity = volume between 50 and 100 cm water tension

water suction to give equal   15–25 cm
     volumes of air and
     water

(b) Arnold Bik (1983) and Boertje (1984):

total porespace	85% (min.)	
air	>25%	determined at pF 1.5 (31.6 cm water tension)
water	>45%	determinations made after applying a pressure of $0.5\,kg\,cm^{-2}$ to the media
shrinkage	<30%	shrinkage = volume lost after drying

Often recommendations for the preparation of mixes include the use of a specific grade or volume of an amendment in order to obtain good air–water relations. Although a certain formula may have achieved the desired results in one locality, it does not follow that this would apply universally. Physical properties of potting mixes are the product of interactions between *all* the materials used and there will be considerable local variation in materials. For example, variation in the particle sizes of peats can be due to the amount of decomposition or to the process used for its extraction, i.e. milling rather than block cutting; particles of river sand are smooth but crushed sands are sharp and angular. Such differences will be reflected in the pore sizes and air–water relations in the media.

For these reasons, the principles of formulating mixes are now considered rather than the descriptions of specific formulae. Results from several investigations (Joiner & Conover 1965, White & Mastalerz 1966, Waters *et al.* 1970, Haynes & Goh 1978) have shown that by varying the ratios and grades of ingredients, mixes having satisfactory physical properties can be made using a wide range of materials.

3.5.1   Physical principles of regulating air–water relations

A theoretical approach to the physical effects of adding amendments to a monodisperse system, i.e. one with a narrow range of particle sizes, was made by Spomer (1974). River sand with particle size less than 1 mm was added in increasing proportions to a compacted powdered silty clay loam. The porosity of the mixtures decreased progressively as the amount of amendment was increased, until a threshold proportion was reached; after this point the porosity of the mixtures increased as the ratio of sand : soil was further increased. At the threshold proportion the small particles of soil were filling or 'clogging' the larger pores of the amendment, i.e. the volume of the mixture was then much less than the volumes of the two separate components.

Air-filled porosity   With materials that have a very narrow range of particle sizes, e.g. sand or bark that has been sieved or graded, it is possible to predict the volume of large pores which create the AFP. If the particles are small, the pores will be small and hence the AFP will be low; when the particle size reaches a critical diameter the pore size and the AFP will increase dramatically. For example, Handreck (1983) found the AFP of sand particles with diameters between 0.25 and 0.5 mm to be only 2% when the moisture tension was 10 cm of water. When the particle size was increased to 0.5–1.0 mm the AFP rose to 35%, at the same moisture tension; the total pore space was virtually similar for both grades. The coarse particles had therefore created fewer but larger pores. Similar results were obtained using graded particles of bark.

However, most potting media are composed of polydispersed materials having a range of particle sizes. Because the shape, texture and internal porosity of the particles vary as well as their size, it is usually accepted that the AFP of a mix cannot be predicted with accuracy, it must be determined empirically. The effects on the total and air-filled porosity of increasing the proportion of either fine sand or coarse grit in an admixture with peat, are shown in Figure 3.8 (Bunt 1984). Whereas, with both minerals there was a negative relationship between the total porosity and the volume of mineral in the mix, only with the fine sand was there a decrease in the AFP, at rates above 25% by volume coarse grit gave a progressive increase in the AFP. The absence of any effect on the AFP when the grit was used at the 25% rate was, in part, due to two opposing effects. Although adding sufficient coarse particles of an amendment will increase the AFP of a mix, the grit with its high BD was simultaneously compressing the mix and thereby reducing some of the natural AFP of the peat. Working with mixtures of

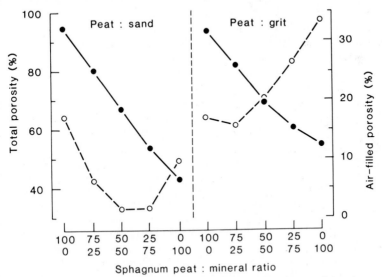

Figure 3.8 Effects of increasing the proportions of fine sand or coarse grit in mixes on the total and air-filled porosities (after Bunt 1984).
Key: ● total porosity; ○ air-filled porosity.

pine bark, sand and brown coal, Richards *et al.* (1986) found that the AFP was related to the percentage of particles greater than 1.0 mm diameter; this they termed the 'coarseness index'. Whether this relationship would also apply to a wider range of media and mixes is not known.

Water retention and availability   The total amount of water retained in mixtures of peat and a mineral, i.e. vermiculite, perlite, calcined clay or sand, depends upon the bulk density of the mix and also the grade of the mineral. For example, the volume of water retained by mixtures having a BD of 0.15 g cm^{-3}, when placed in containers 10 cm deep, was 85% for mixes made with 'fine' grades and 75% for those with 'coarse' grades of minerals. Increasing the BD to 0.8 g cm^{-3} decreased water retention; mixes with 'fine' grades then had 66% and those with 'coarse' grades only 45% water (Bunt 1984).

The tension required to remove water from a potting medium, and the pore size distribution of the medium, can only be known from desorption curves. Examples of the effects of adding minerals having different particle gradings and bulk densities to a sphagnum peat are shown in Figures 3.9 and 3.10. When the added mineral had a high BD (sand or grit), the volume of water released at low tensions decreased as the proportion of fine particles in the mix was increased, i.e. aeration at low tensions was reduced. Coarse particles had the reverse effect of releasing more water, thereby improving aeration. These minerals also reduced the total porosity in proportion to their ratio in the mix (see Fig. 3.8). Porous minerals with a low BD (e.g. vermiculite) had little effect on the total porosity, and the desorption curve was also less affected by the proportion of vermiculite in the mix (see Fig. 3.10). Porous amendments such as vermiculite and bark hold a reserve of water within their small pores; however, this will only be released at much higher tensions than normally occur with container grown plants in greenhouses.

3.5.2   *Water-absorbing polymers*

Gel-forming synthetic polymers as used to aid water retention in sandy soils are sometimes used in potting media. The polymers can be of three types, i.e. starch copolymers, polyvinylalcohols and polyacrylamides. These materials can absorb relatively large quantities of water, usually in the region of 200–400 times their weight of deionized water, but considerably less if the water contains nutrients or salts. Depending upon the type of gel and the salt, water absorption can be reduced by 60–80%. Solutions containing divalent ions, e.g. Ca^{2+}, Mg^{2+}, SO_4^{2-}, reduce absorption more than solutions containing monovalent ions, e.g. Na^+, HCO_3^-, Cl^- (Johnson 1984). James & Richards (personal communication) have also found that in the presence of iron sources producing hydrated ionic species water absorption was reduced by 50%, chelated iron was less detrimental than other iron sources.

Recommended rates of use range from 1–8 g l^{-1} of mix; in some cases prewetting the granules is recommended before they are added to the

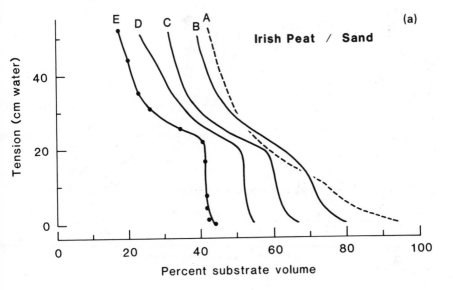

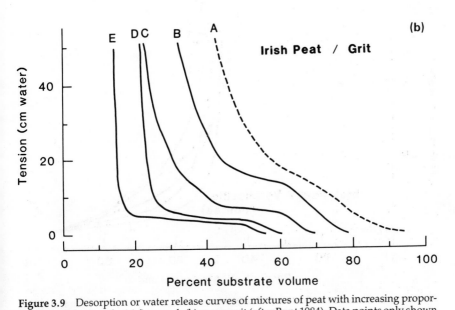

Figure 3.9 Desorption or water release curves of mixtures of peat with increasing proportions of solid minerals: (a) fine sand, (b) coarse grit (after Bunt 1984). Data points only shown for the fine sand treatment (E).
Key: A 100% peat; B 75% peat 25% mineral; C 50% peat 50% mineral; D 25% peat 75% mineral; E 100% mineral.

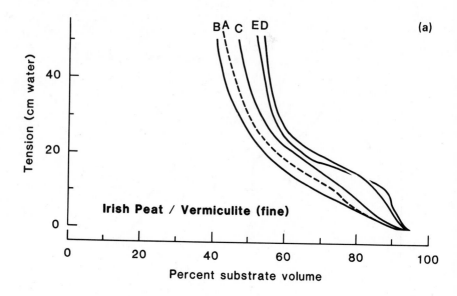

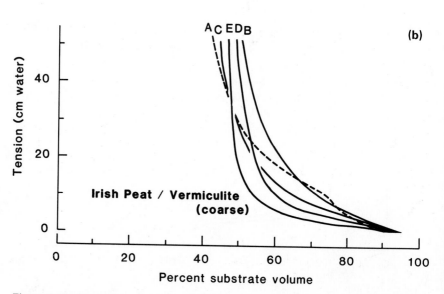

Figure 3.10 Desorption or water release curves of mixtures of peat with increasing proportions of a porous lightweight mineral: (a) fine vermiculite, (b) coarse vermiculite (after Bunt 1984).
Key: A 100% peat; B 75% peat 25% vermiculite; C 50% peat 50% vermiculite; D 25% peat 75% vermiculite; E 100% vermiculite.

mix, this reduces the risk of the granules swelling in the container and thereby reducing aeration. The reported benefits from their use include a reduction in the frequency of watering (Still 1976) and an increase in the shelf life or time to wilting (Gehring & Lewis, 1980).

The effect which polymers have on the water absorption and release characteristics varies with the medium; also, to obtain the correct interpretation, the basis on which the results are being presented should be fully appreciated. About 90% of the water held by the polymer is released at tensions between 0.1 and 2 bars, and usually peat with polymer added releases 3–4% less water at low tensions, i.e. below 100 cm of water. Unpublished work by the writer shows that polymers increase the available water content of sands but have little effect on peat. Available water is defined as the amount per unit *volume* after drainage, less the amount still present at the permanent wilting point. Treated and untreated peat mixes had similar volumes of available water, i.e. about 65%; when added to a coarse sand, however, the polymer increased the volume of available water from 18.5% to 32.8%. Polymers expand when fully hydrated and so reduce the BD of the mix by 25–30% in comparison with non-treated mixes. If the results are expressed on a *weight* basis, rather than the volume basis, the polymer treatment gives an apparent increase of available water equal to 24% for peat and 240% for the coarse sand. In container growing, polymers could give moderate increases in the volume of available water if the medium has coarse particles with low water retention, e.g. sands and pine bark. However, with sphagnum peat the effect will be small.

Prewetting the polymer granules before mixing ensures the maximum water absorption and therefore a low bulk density for the mix. However, it has been observed that when plants are growing in mixes made with the prewetted polymer, granules are often incompletely expanded, even after watering. Failure of the polymer granules to re-expand to their former volume is due to a greater salinity of the soil solution compared to the water originally used to prewet the granules. The incomplete re-expansion of the granules leads to the development of air pockets around the polymer, thereby improving aeration. Thus, the improved root growth sometimes seen in containers of overwintered plants grown in a medium with added polymer is an effect of improved aeration rather than increased available water content.

3.6   WETTABILITY OF MIXES

It is common experience that mixes made with materials such as peat and pine bark can be difficult to wet initially, and also to rewet after they have dried-out in the container. The wetting resistance of peat has been attributed to iron humates and a strongly adsorbed air film; other materials have resins and waxes which make wetting difficult. The most convenient way of wetting peat is to add water through a spray line when the mixer is working. However, it is essential that the fertilizers are mixed with the peat while it is still dry. If powdered fertilizers are added to wet peat their distribution is very uneven, and this can result in large variations in plant growth. Granular fertilizers are not as susceptible to this problem.

The moisture content of peat in bales is usually between 45% and 65% on the, normally quoted, moist weight basis. For good potting conditions, and to enable the peat to take up water readily, the moisture content should be increased to about 70–75%. The efficiency or ease with which hydrophobic materials can be wetted is proportional to the amount of water that they already contain. Airhart *et al.* (1978b) reported that when the moisture content of milled pine bark was less than 35% on the moist weight basis (c. 55% by dry weight) very little of the water applied by misting was retained by the bark. As the moisture content of the samples was increased from 35% to 50%, they became progressively easier to wet. This difficulty with the first wetting of dry bark in containers is the reason why plants have sometimes shown an initial delay in growth and establishment. Other tests showed that although a peat–vermiculite mix was 78% saturated after three days of frequent misting, dry bark was only 20% saturated after ten days, and 70% saturated after 48 days.

There are two main ways in which the wettability of potting mixes can be improved, i.e. by the incorporation of minerals, or by addition of a surfactant to lower the water surface tension.

3.6.1 Mineral additives

Most minerals can be easily wetted, although some sands and mineral soils are naturally water repellant, e.g. those found in parts of California. One of the reasons for the early GCRI recommendation to include sand with peat was the improved ability of the mix to absorb water in the dry state. Beardsell & Nichols (1982) showed that coarse sand retained its wettability irrespective of the number of days which had elapsed between irrigations. This ability to rewet is quantitatively transferred to container mixes in proportion to the amount of sand used. They concluded that to obtain a good degree of wettability fresh pine bark should have at least 30% by volume of coarse sand mixed with it. Difficulties in rewetting media were attributed to two causes, materials may be hydrophobic, e.g. pine bark, or exhibit shrinkage on drying, e.g. brown coal (lignite). Mixing of materials can result in the substrate having a lower wettability than the materials themselves. For example, the rewettability of pine bark was 83.5% and brown coal 76.2%; however, a mixture of 1 part pine bark and 2 parts brown coal had a wettability of only 55.9%. Both vermiculite and perlite improve the wettability of mixes.

3.6.2 Surfactants (wetting agents)

Another approach to the problem is to reduce the surface tension of the water, thereby enabling it to spread and penetrate the mix more readily. When water comes into contact with a hydrophobic surface it forms small beads, the angle between the surface and the side of the bead (the contact angle) is large and indicates the antagonism between the water and the surface of the material. If a surfactant is added to the water the surface tension is reduced, thereby decreasing the contact angle.

It is essential that the surfactant is not toxic to plants. Non-ionic surfactants are chemically less active, but also less phytotoxic, than either the anionic or cationic surfactants. Sheldrake & Matkin (1971) tested more than 30 wetting agents and found some of the most efficient for wetting air-dry peat were also phytotoxic. Parr & Norman (1964) found that, in general, wetting agents based on polyethylene sorbitan fatty acid esters were less inhibitory to roots than those with ether or ether-alcohol structures. It is also important that the wetting agent does not degrade quickly in the substrate. Valoras et al. (1976) found that sphagnum peat absorbed wetting agents at a much higher rate than did a sandy loam; also, degradation of surfactants in the peat was much slower than in the sandy loam. After 280 days less than 30% of the wetting agent in the peat had decomposed, whereas in mineral soils decomposition ranged from 70–90%.

Examples of surfactants available in the USA are 'Aqua Gro', 'Triton X-100' and 'SurfSide'; and in the UK: 'Aqua Gro', 'Nonidet LE' and 'Agrol'. They are usually used at a concentration of 0.1% or 1000 p.p.m. (1 fl. oz to 6 gal.) but with materials that are very hydrophobic, e.g. uncomposted pine bark, twice this concentration may be required. Wetters can also be included when the mix is being prepared. Either 100–150 cm^3 is dissolved in 20 l of warm water which is then sprayed over 1 m^3 of medium while it is still in the mixer, or 100–150 cm^3 of the concentrated wetter is sprayed onto 1 l of fine, dry vermiculite, which acts as a carrier.

Figure 3.6 shows the AFP values of a hydrophobic peat, plus or minus surfactants. Water content values calculated from the data showed that for peat alone, after 40 days of being watered twice daily, the water contents were 64% and 81% (corresponding to AFP values of 30% and 13%). For peat treated with surfactant, the water contents were 83% and 89%. After 70 days of twice daily irrigation the wettability of the untreated peat had increased only slightly.

By reducing the surface tension of the water, surfactants can in some situations improve drainage and thereby aeration.

3.7 COMPACTION

Often irrigation and natural settling will result in a slight increase in the BD of potting mixes, especially with some peats when the containers have been loosely filled. More positive compaction occurs from mechanical or hand compression during potting. Traditionally, chrysanthemums were potted firmly using a wooden, wedge-shaped rammer, the purpose being to reduce the growth rate and make the flowers more resistant to damage in handling. However, modern practice is to minimize compaction and obtain rapid growth.

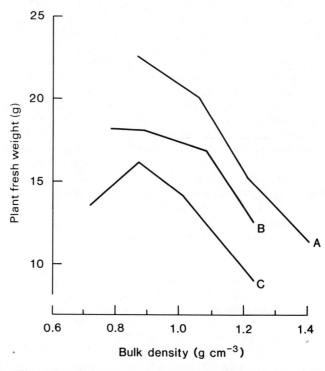

Figure 3.11 Effects of media compaction on the growth of tomato vary with the structure of the soil used in the mix. All mixes made with 60% mineral soil, 25% peat and 15% grit. The tests were made on three separate occasions, comparisons between the three media cannot be made (redrawn from Bunt 1961).
Key: A a 'fine sand' soil with poor structure; B soil with an intermediate structure; C a 'brick earth', well aggregated.

Compaction has several effects on the biological and physical properties of mixes. Total pore space is reduced, but more importantly so is the volume of large pores which controls aeration and drainage. Moderate compaction can increase the amount of available water, but higher rates of compaction will reduce it. Other effects are: an increase in the salt concentration because of the extra amount of medium, and thereby fertilizer, per unit volume; restriction of root penetration because of the small size and rigidity of the pores; and reduction of the rate of mineralization of organic nitrogen.

Compaction often causes a reduction in vegetative growth, but this effect varies with the medium. For example, Figure 3.11 shows the fresh weights of tomato plants grown in composts made from soils of different textures and structures. Each compost had 60% by volume of soil, 25% sphagnum peat and 15% coarse grit. Compost made with a well-aggregated soil (C) had a permeability of 262 cm^3 min^{-1}, and moderate compression increased plant growth. Compost made with a poor structural soil (A) had a permeability of 94 cm^3 min^{-1} and all degrees of compression adversely affected plant growth. Compost made with soil (B) had a permeability of 177 cm^3 min^{-1} and results were intermediate to the other two composts. High rates of compaction, i.e. when loose compost was compressed to 70% or 60% of its original volume, adversely affected plant growth in all composts. Results obtained by Flocker et al. (1959) also showed that the growth response to compaction varied with the soil type.

CHAPTER FOUR

Principles of nutrition

The main requirements for plant growth are adequate levels of light, temperature and carbon dioxide in the aerial environment, and of water, air and mineral nutrients within the growing medium. It is the various aspects of mineral nutrition with which we are now concerned.

The leaf and stem tissue of a typical pot plant can be regarded as consisting of 90% water and 10% solids; these values will vary a little with the species, the general environment and the age of the plant. The dry matter can next be subdivided into organic and inorganic fractions consisting of approximately 90% organic compounds such as cellulose, sugars, proteins etc., with the remaining 10% being minerals such as calcium, potassium, magnesium, phosphorus and microelements. The amount of dry matter produced by a pot plant and hence the total amount of minerals required, varies considerably, ranging from about 30 g for a pot chrysanthemum grown in a 14 cm pot holding 1 l of medium, to 3 g for a *Saintpaulia* grown in a 8 cm pot holding 300 cm^3 of medium; the dry weight of a box of bedding plants grown in 4.5 l of medium would be about 20 g. Mineral composition of the dry matter also varies slightly with the plant species and the nutrition it receives, but the following would be typical proportions of minerals found in the leaves of the average pot plant:

nitrogen 3–4.5%, phosphorus 0.3–0.6%, potassium 3–4.5%, calcium 1–2%, magnesium 0.2–0.5%

Microelements such as copper and boron are present in much smaller concentrations, e.g. 10–100 p.p.m. Although the minerals represent only about 1% of the total fresh weight of the plant, they are nevertheless essential if the various growth processes are to function efficiently. Adequate levels of minerals must therefore be maintained in the medium at all times if maximum rates of growth are to be achieved.

4.1 CATION EXCHANGE CAPACITY

Plant nutrients are normally applied as salts, which are composed of atoms carrying electrical charges known as ions; those with a positive charge are called *cations* whilst those with a negative charge are called

64

anions. The salt potassium nitrate (KNO_3) comprises potassium (K^+) cations and nitrate (NO_3^-) anions.

One of the important mechanisms which help to regulate the supply of certain nutrients to the plant is known as the cation, or base, exchange capacity (CEC). This is defined as the sum of the exchangeable cations or bases that a soil can absorb per unit weight and is usually expressed as milligram equivalents per 100 g; this is sometimes abbreviated to me or meq $100\,g^{-1}$. Cation exchange capacity is normally associated with the clay particles in a mineral soil, and although organic matter such as humus, peat and the roots of plants, and other minerals such as vermiculite, also have cation exchange capacities, it is convenient for the present to follow the convention of considering this phenomenon in relation to mineral soils. Each clay particle has a number of negative charges on its surface and when fertilizers are applied to the soil the cations, such as Ca^{2+}, Mg^{2+} and H^+, are attracted to the clay particle. The cation exchange process illustrated in Figure 4.1 shows the reaction which occurs when lime, in the form of calcium carbonate or ground lime, is applied to an acid soil; this process also occurs with other salts or fertilizers. The cations commonly associated with plant nutrition are Ca^{2+}, Mg^{2+}, K^+, NH_4^+ and Na^+ – these have been arranged in order of decreasing retention by the clay particle, i.e. sodium and ammonium are not as strongly retained as magnesium and calcium. Traditionally the CEC of mixes and their separate ingredients have been determined using ammonium acetate at pH 7. For loam-based mixes, which are normally used within the pH range 6.2–6.8, a CEC determination made at pH 7 is a good reference value. However, such a determination is not so useful for many peat and lightweight mixes in which plants are grown at pH values of between 5.0 and 5.5. This is because the CEC of materials is known to be pH dependent, organic matter being more affected than clay. For example, Helling *et al.* (1964) showed that the exchange capacity of organic matter increased by 140 meq $100\,g^{-1}$ when the pH was raised from 3.5 to 8.0, whereas for clay the increase was only 18 meq $100\,g^{-1}$. The importance of determining the exchange capacity of potting media at pH values similar to those at which the plants are grown is demonstrated by the results of Haynes (1982) shown in Table 4.1. Several methods were compared using a New Zealand sedge peat, H4 on the von Post scale, which had been limed to give three pH values. The exchange capacities determined by calculation and by ammonium chloride were much lower than those using methods conventionally employed for mineral soils. The calculated and ammonium chloride exchange capacities also increased with the pH of the peat, whereas conventional methods gave a constant exchange capacity irrespective of the pH. The CECs of peats and bark largely indicate the potential amounts of exchangeable divalent ions, i.e. Ca^{2+} and Mg^{2+}; most of the monovalent cations NH_4^+ and K^+ will be water-soluble.

The cation exchange capacities of some materials and potting mixes is given in Table 4.2. Although plants can be grown satisfactorily in a wide range of materials irrespective of their cation exchange capacities, management is usually easier when the medium has a reasonable exchange capacity.

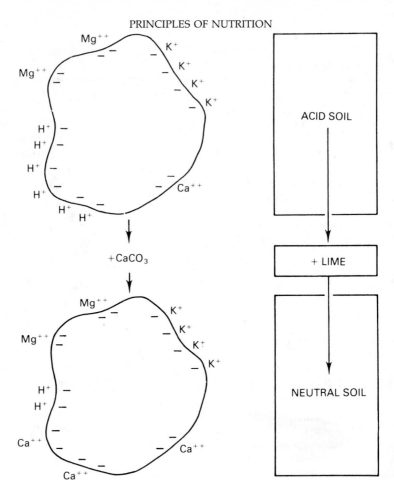

Figure 4.1 Diagrammatic representation of a clay particle showing the cation exchange process when calcium carbonate is applied to the soil.

Table 4.1 CEC of a sedge peat at three pH values determined by four methods (from Haynes 1982).

| pH | | CEC (meq $100\,g^{-1}$) | | |
	effective*	NH_4Cl	NH_4OAc†	$BaCl_2$-TEA‡
4.3	21.6	36.9	107.2	181.1
5.3	57.9	64.7	109.7	179.2
5.6	72.4	73.6	110.9	178.9

* Effective CEC was calculated from the sum of the exchangeable bases plus exchange acidity.
† $1\,M\,NH_4OAc = pH\,7.0$.
‡ $BaCl_2$-TEA = pH 8.2.

Table 4.2 Cation exchange capacities of some materials and mixes.

Materials and mixes	Cation exchange capacity	
	meq $100\,g^{-1}$	meq $100\,cm^{-3}$
[1] humus	200	
vermiculite	150	
montmorillonite	100	
illite	30	
kaolinite	10	
[2] fine clay	56–63	
coarse clay	22–52	
silt	3–7	
[3] perlite	1.5	
[4] peat, sphagnum	100–120	
mixes (% by volume)		
[5] 50% sphagnum peat, 50% sand	8.01	5.35
75% sphagnum peat, 25% sand	18.16	6.75
50% sedge peat, 50% sand	21.11	17.12
75% sedge peat, 25% sand	40.05	21.80
John Innes compost	8.8	7.7
[6] 25% peat, 75% perlite	11.2	1.0
50% vermiculite, 50% peat	140.96	32
66% pine bark, 33% perlite	23.58	5

[1] Thompson (1957); [2] Whitt & Baver (1930); [3] Morrison *et al.* (1960); [4] Puustjärvi (1968); [5] Bunt & Adams (1966a); [6] Joiner & Conover (1965).

4.2   ANION EXCHANGE CAPACITY

In addition to the cations, plants also require the negatively charged anions, such as nitrate (NO_3^-), chloride (Cl^-), sulphate (SO_4^{2-}) and phosphate ($H_2PO_4^-$). Although mineral soils have a permanent negative charge and so attract cations as decribed (§ 4.1), they do not possess a similar attraction for anions. Some phosphate anions are, however, adsorbed by the clay particles and can act as replaceable anions. The pH or acidity of the soil has a large effect on the forms and availability of phosphorus to the plant. When a water soluble form of phosphorus such as superphosphate is added to a mineral soil, it is quickly converted from the soluble monocalcium phosphate to the insoluble dicalcium phosphate. Under strongly acid conditions, the phosphorus combines with aluminium and iron to give strongly fixed or insoluble aluminium and ferric phosphates. These insoluble phosphates become slowly available to the plants by the action of the weak acids present in the soil solution. Mineral soils usually contain only very low concentrations of phosphorus in a water-soluble form and this would be quickly depleted by the plant, unless it was continuously renewed from the insoluble or reserve forms of phosphorus present in the soil. It has been estimated that the amount of phosphorus released in this manner is equivalent to the complete replacement of the water-soluble phosphorus several times each day.

Unlike the mineral soils, peats do not have a significant anion exchange capacity and they contain relatively little iron and aluminium. A large part of the phosphorus applied as superphosphate to the peat therefore remains in the water-soluble form and is readily available to plants; however, two disadvantages are that it can be easily leached and it can be toxic to some species (§ 6.1.4). The amount of phosphorus that remains in water-soluble form largely depends upon the amount of calcium carbonate added to the peat. For example, when a solution containing 120 mg l^{-1} of water-soluble phosphorus (from superphosphate) was added to a sphagnum peat that had been limed to give three pH values, the percentages of water-soluble phosphorus were:

pH of peat	4.9	5.8	6.3
% water-soluble phosphorus	71	67	56

4.3   AVAILABILITY OF NUTRIENTS: LOAM v. LOAMLESS MIXES

4.3.1   Nutrient availability

The difference in the behaviour of the major plant nutrients when applied to a peat–sand mix and a mineral soil compost (John Innes) is shown in Table 4.3. The nitrogen, phosphorus and potassium levels in these two media were determined both before and after adding a base fertilizer which supplied known amounts of ammonium, nitrate, phosphorus and potassium; the peat–sand mix also had the normal amount of calcium carbonate and Dolomite limestone added. The availability of these nutrients was then assessed by keeping the media moist but avoiding leaching for two weeks to allow any fixation of the nutrients to occur. Each mix was then extracted with two extractants, viz. distilled water and Morgan's solvent, the latter being 0.52 N acetic acid buffered with 0.73 N sodium acetate and having a pH of 4.8. Nutrients extracted by this and

Table 4.3   Percentage of added nutrients found by chemical analysis of a mineral soil compost (John Innes) and a peat–sand mix (based on extracts using water or Morgan's solvent).

Nutrient	John Innes		Peat–sand	
	water	Morgan's	water	Morgan's
ammonium nitrogen	69	88	79	90
nitrate nitrogen	81	90	87	95
phosphorus	9	16	56	66
potassium	40	57	74	91

similar types of weak acid extractions are generally regarded as being immediately available to the plant.

With the John Innes compost, 16% of the added phosphorus was extracted by the weak acid and only 9% by the water, whereas in the peat–sand mix the corresponding values were 66% and 56%. The weak acid had extracted 80% more phosphorus than the water had from the John Innes compost but only 20% more than the water had from the peat–sand mix. The much greater water-solubility of phosphorus in peat mixes means that it is less likely to become a limiting factor in plant nutrition, *providing that it is not lost by leaching.* Potassium was also more easily removed by both extractants from the peat–sand mix than it was from the John Innes compost.

4.3.2 Loss by leaching

The relatively high frequency of irrigation required by container-grown plants has already been discussed. Inevitably, this must result in some loss of water from the containers, together with the possible loss of nutrients. In view of the high solubility of most minerals in peat mixes, as already shown in Table 4.3, the rate at which minerals could be lost from loamless mixes made from different materials was examined under typical growing conditions (Bunt 1974a). Two sets of mixes were prepared, one consisting of 75% by volume of sphagnum peat and 25% of fine sand, the other of equal parts of sphagnum peat and vermiculite. Known amounts of nutrients were added to individual pots each holding one litre of medium. Sufficient water was then given at a high rate of application, i.e. equivalent to hose watering, to collect leachate in increments of 50, 75, 125, 250 and 500 cm^3 thereby making a total of one litre of leachate from each pot. This was equivalent to the total volume of each pot, or an irrigation equivalent to 6.5 cm depth of water.

The total amounts of nutrients recovered from these mixes are given in Table 4.4 and the rates at which they were leached are

Table 4.4 Percentage of added nutrients recovered in one litre of leachate from peat–sand (3:1) and peat–vermiculite (1:1) mixes, in 1 litre pots.

Nutrient	Peat–sand	Peat–vermiculite
ammonium nitrogen	81	33
nitrate nitrogen	87	75
phosphorus	60	43
potassium	70	45

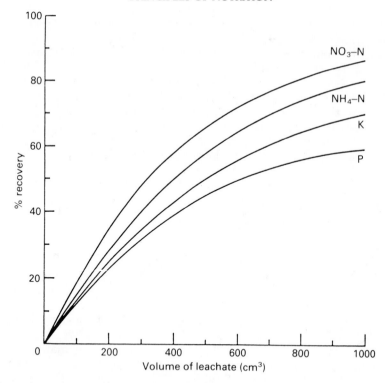

Figure 4.2 The rate at which nutrients are leached from a peat–sand mix (compare with Fig. 4.3).

shown in Figures 4.2 and 4.3. All nutrients showed an exponential rate of loss, with 28% or more of the nutrients added to the peat–sand mix being lost in the first 250 cm^3 of leachate. Apart from nitrate nitrogen, the peat–vermiculite mix showed a much lower rate of nutrient loss, only 15–22% of the nutrients being lost in the first 250 cm^3 of leachate. In addition to its known ability to hold potassium and ammonium in an exchangeable form, the vermiculite also reduced the amount of phosphorus lost by leaching.

4.4   CHEMICAL ANALYSIS OF LIGHTWEIGHT MEDIA

The physical and chemical characteristics of peat and other lightweight mixes differ appreciably from those of loam-based mixes. Consequently many of the methods traditionally used in preparing samples for chemical analysis are not considered satisfactory for use with modern growing media. Various methods that can be used to analyze media are summarized in Figure 4.4. The principal factor in deciding whether to use a suspension method or saturated media extracts will be the number of samples to be analyzed. For advisory or extension purposes where large

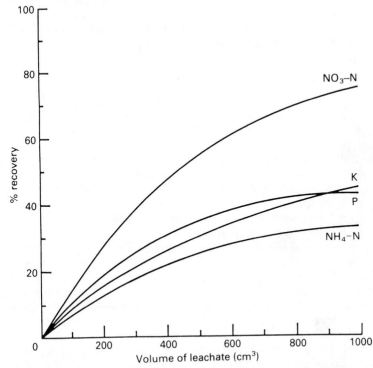

Figure 4.3 Inclusion of vermiculite in the mix significantly reduced the rate of loss of nutrients, apart from nitrate nitrogen.

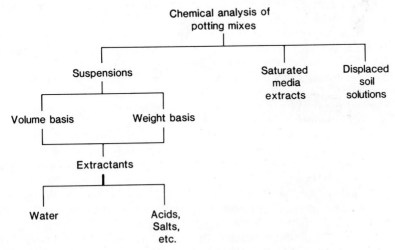

Figure 4.4 Alternative procedures commonly used for the chemical analysis of potting mixes.

numbers of samples must be analyzed quickly, a suspension method is often preferred. For research purposes, where there are fewer samples and a more fundamental interpretation of the nutrient–plant relationship is required, the saturated medium extract method is preferred. The main points with the various methods of analysis are summarized below.

4.4.1 Suspensions

With suspensions, the first decision must be whether the sample is to be air-dried, ground and then sieved before it is extracted with a solvent, or whether it is to be analyzed in the moist, or fresh, state.

Drying This can result in the separation of ingredients of different density, e.g. peat and sand, thereby resulting in sampling errors. Organic materials such as peat and bark are difficult to rewet after drying. Also, drying reduces the amounts of nitrate nitrogen, phosphorus, potassium, calcium and magnesium that are extracted, whereas the amount of ammonium nitrogen can be increased.

Grinding This changes the media structure and increases its bulk density. It also creates problems if slow-release fertilizers such as Osmocote, Nutricote, Ficote, MagAmp or Enmag are used. Grinding breaks up the fertilizer particles and thereby gives much higher values of apparently available nutrients.

Sampling This can either be on a weight or on a volume basis. Whereas the accuracy of measuring media by weight is much greater than by volume, the interpretation of a volume-based analysis is much easier. It automatically compensates for the bulk density of the mix, which can vary from $0.1–1.2$ g cm^{-3}. As fertilizers are added to a *volume* and not a *weight* of medium, large errors can occur in the interpretation of analyses based on extraction by weight if corrections for the differing bulk densities are not applied. Differences in the moisture content of mixes can also result in large errors if sampling is by weight.

For these reasons sampling by volume is preferred, but it is necessary to use relatively large volumes, i.e. at least 50 cm^3 and preferably 100 cm^3. It is also necessary to adopt a standard procedure for filling the measure and so avoid errors by differences in packing.

Extractants Whereas with mineral soils weak acids or salt solutions are normally used to estimate water-soluble and exchangeable nutrients, with modern lightweight mixes a large proportion of the major plant nutrients are soluble (see Table 4.3 & Fig. 4.2) and water is the preferred extractant. The contrast between the relationship of the plant with its nutrient supply when growing under field conditions (i.e. in a mineral soil, with a relatively low root density) and when grown in containers of lightweight media at high root densities, is shown in Figure 4.5. The differences in the areas of the solid and liquid phases shown in the two

Nutrition in mineral soil - Field conditions

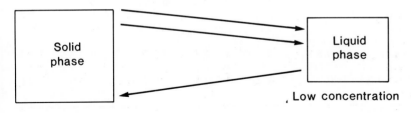

Low root density

Diffusion – mass flow

Solid phase → Liquid phase

Low concentration

Exchangeable nutrients
extracted with
weak acids and salts

(a)

Nutrition in greenhouse media

High root density

Root interception – mass flow (liquid feeding)

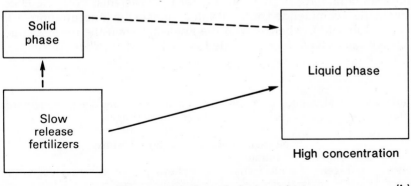

Solid phase

Slow release fertilizers

Liquid phase

High concentration

Water extraction of plant nutrients   **(b)**

Figure 4.5 Contrasts in the relationship between the plant and its nutrient supply; (a) field crops with low root density in mineral soils, (b) plants with high root density in containers of lightweight media.

systems of growing is intended to convey their relative importance in the nutrition of plants; the reason for the choice of water as the extractant for soil-less media is clear.

Methods of chemical analysis There are two methods of suspension analysis in general use in western Europe, both are based on a water

Table 4.5 Standards for nitrogen, phosphorus, potassium and magnesium in the 1:1.5 volume extract (Dutch method).

	Low	Fairly low	Normal	Fairly high	High	Very high
nitrogen $(meq\,l^{-1})$	<1.9	1.9–3.6	3.7–5.4	5.5–7.2	7.3–9.0	<9.0
phosphorus $(mg\,l^{-1})$	<8	8–14	15–21	22–28	29–35	<35
potassium $(meq\,l^{-1})$	<0.8	0.8–1.4	1.5–2.1	2.2–2.8	2.9–3.5	<3.5
magnesium $(meq\,l^{-1})$	<0.7	0.7–1.2	1.3–1.8	1.9–2.4	2.5–3.0	<3.0

extract of a volume sample. The Dutch method (Sonneveld *et al.* 1974) uses a narrow ratio of one volume of medium to 1.5 volumes of distilled water. To avoid errors due to variations in the quantity of water already present in the media, the water content is adjusted to pF 1.5 before extraction. With the English method (Johnson 1980) one volume of medium is extracted with six volumes of distilled water; because of the much wider extraction ratio no allowance is required for variation in the water content of the samples. Guidelines for the desirable nutrient concentrations have been published for both systems. With the Dutch system the recommended nutrient values refer to the concentrations *in the extract* (Table 4.5), whereas with the English system the recommended nutrient values have been calculated as $mg\,l^{-1}$ *in the medium* (Tables 4.6 & 4.7).

Table 4.6 Standards for ammonium nitrogen, nitrate nitrogen, phosphorus, potassium and magnesium in the 1:6 volume extract (English ADAS method).

Index	Nutrient concentration ($mg\,l^{-1}$ of mix)				
	ammonium nitrogen	nitrate nitrogen	phosphorus	potassium	magnesium
0	<20	<15	<4	<25	<5
1	21–50	16–25	5–7	26–50	6–10
2	51–100	26–50	8–11	51–100	11–15
3	101–150	51–80	12–18	101–175	16–25
4	151–200	81–130	19–28	176–250	26–35
5	>200	131–200	29–40	251–400	36–50
6		201–300	41–55	401–650	51–85
7		>300	56–75	651–1000	86–150
8			76–100	1001–1500	151–200
9			>100	>1500	>200

Ammonium nitrogen values should not exceed index 3 for unused mixes; young plants may be affected at index 4.

Table 4.7 Desirable indices for crops (English ADAS method).

	Nitrogen	Phosphorus	Potassium	Magnesium
seed mixes	1	4	2	2
pot plants and bedding plants	3–4	5	3	4
tomatoes in peat modules	4–5	7–8	5	4

Some laboratories use weak acids or salt solutions as extractants and consequently larger amounts of plant nutrients are removed. The extent to which this occurs depends upon several factors, including the strength and type of extractant, chemical characteristics of the medium, i.e. its exchange capacity and ability to fix phosphorus, and also the length of the extraction time (Wilkerson & O'Rourke 1983). As all extracting solutions and procedures differ in the amounts of major nutrients they remove, it is necessary to make a series of trials in which plant growth and the amount of nutrients present in the leaves are correlated with the amount of nutrients removed from the media; this information forms the basis for diagnosis and manurial recommendations.

4.4.2 Saturated media extract (SME)

In parts of the USA this method of analysis is used for extension or advisory work as well as for research purposes. Although originally introduced for classifying the salt status of field soils there is now considerable experience with its use for potting mixes (Warncke & Krauskopf 1983). With this method just enough water is added to saturate the medium, there should not be any free water present. Once equilibrium is reached a pH reading of the saturated medium is taken, then the solution is extracted by a vacuum filter and is used for the usual chemical analysis. The amount of water in the saturated medium will be about four times that present at the permanent wilting point (pF 4.2) and approximately twice that in a mix that has been watered and drained. The actual relationships depend upon the amount of organic matter in the mix and the depth of the container.

The advantages of this method are:

(a) the amount of water added, and hence the dilution of the nutrients, is automatically adjusted to the texture of the medium, no further adjustments for differences in the bulk densities of the mixes are required;

(b) the concentration of nutrients, the salinity and the pH can be easily related to those experienced by the plant;

(c) in addition to measuring the concentration or *intensity* of the nutrients, the *balance* between nutrients is also easily calculated (Geraldson 1967). With suspension methods of analysis there will be some change in the balance of nutrients as the ratio of media to water is varied, at wide ratios the proportion of monovalent to divalent cations in solution increases.

Table 4.8 Standards for nutrients determined by the saturated media extract method.

	Nutrient concentration (p.p.m.)				
	nitrate nitrogen	phosphorus	potassium	calcium	magnesium
low	<39	<2	<59	<79	<29
acceptable	40–99	3–5	60–149	80–199	30–69
optimum	100–199	6–10	150–249	>200	>70
high	200–299	11–18	250–349		
very high	>300	>19	>350		

Table 4.9 Desirable nutrient balance in saturated media extract.

Nutrient	% of total soluble salt
nitrate nitrogen	8–10
ammonium nitrogen	<3
potassium	11–13
calcium	14–16
magnesium	4–6
sodium	<10
chloride	<10

Disadvantages of this system are:

(a) inexperienced operators can have difficulty in deciding on the point of saturation, this is especially true with materials such as pine bark and fibrous peats;
(b) some laboratories find the method too time consuming.

Guidelines for nutrient concentrations in the extract are given in Table 4.8. Seedlings would generally require to be grown at the lower levels, i.e. in the 'acceptable' to 'optimum' range, whereas vigorous plants such as pot chrysanthemums could be grown in the 'optimum' to 'high' range. The desired balance of nutrients in the saturation extract is given in Table 4.9.

4.5 NUTRIENT UPTAKE BY THE PLANT

Plants absorb nutrients in the form of cations and anions through their root hairs which are covered by a film of water containing dissolved nutrients. This is sometimes referred to as the soil solution. The concentration of nutrients found in plant roots, however, is many times greater than that present in the soil solution and it is evident that diffusion cannot play an important part in nutrient absorption. Nutrient absorption is an energy consuming process; sugars that have been manufactured in the leaves are transported to the roots where they are used to provide the energy required for root respiration and nutrient uptake. Ions are absorbed selectively and independently by the plant. For example, if a fertilizer such as ammonium sulphate is added to the medium the plant will absorb many more NH_4^+ ions than SO_4^{2-} ions; similarly when calcium nitrate is added, more NO_3^- ions will be absorbed than Ca^{2+} ions. The rate of ion absorption is dependent upon several factors other than their actual concentration in the soil solution. A low supply of sugars in the plant, low soil temperature and low oxygen supply will all reduce the rate of absorption. Plants also release ions through their roots; for example, H^+ ions can be released in exchange for Ca^{2+} ions, and HCO_3^- ions can be released in exchange for NO_3^- ions.

Frequently, the rate of plant growth can be increased by increasing the supply of nutrients in the mix, but it is essential that a balance between the nutrients is maintained at all times. If, for example, the amount of nitrogen, phosphorus and potassium available to the plant is increased, it is quite possible that magnesium deficiency symptoms will appear. This will have been caused partly by the higher growth rate and the greater demand for magnesium, and partly by the antagonism which exists between potassium and magnesium; high potassium levels actually render magnesium less available. This is one of the best known instances of antagonism in plant nutrition but several other examples exist. The reverse case of potassium deficiency in chrysanthemums, induced by the use of fertilizers very high in magnesium, has been observed in experiments with loamless mixes made at the Glasshouse Crops Research Institute. Some ions, however, increase the absorption of others, e.g. the anion NO_3^- is more effective than other anions in increasing the uptake of the cations Ca^{2+} and K^+; the NO_3^- anion also reduces the uptake of the SO_4^{2-} anion. Some minerals will also partially substitute for others in the plant: sodium, for example, will reduce the need for potassium.

Although an increase in the general supply of nutrients will often give an increased rate of growth, it is possible to arrive at a situation whereby growth has been depressed by too great a supply of nutrients. This change in the response of the plants to the supply of nutrients is represented diagrammatically in Figure 4.6. Usually the plant does not absorb nutrients at a uniform rate over its growing period: when plants are young and growing actively they often contain higher concentrations of nutrients than older, slower growing plants. A typical pattern of dry matter accumulation and nutrient uptake is shown in Figure 4.7. At four weeks after potting, the chrysanthemums had made 16% of their final dry weight but had already absorbed 35% of the final phosphorus uptake. Also, nutrients are not always distributed evenly within the plant. Plants grown in mixes with high levels of copper have been found to have much higher concentrations of copper in their roots than in their shoots.

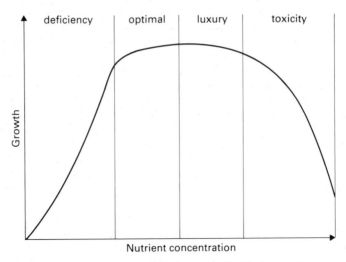

Figure 4.6 The relationship of the rate of plant growth with the nutrient concentration in the medium. Once the optimal nutrient level is reached, there is the risk of suppressing growth.

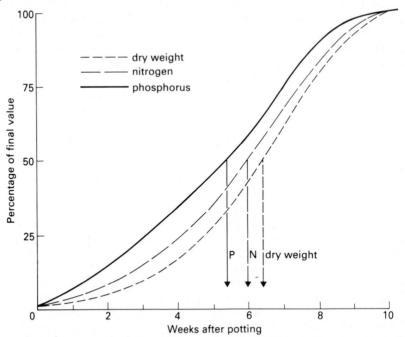

Figure 4.7 Typical curves showing the rate of dry weight accumulation and nutrient uptake for pot chrysanthemums. Phosphorus was taken up at a high rate by the young plants.

4.6   ACIDITY (pH)

Materials used in potting mixes differ widely in their pH (acidity); peats are usually acid in reaction and some of the vermiculites are alkaline. The degree of acidity or alkalinity is customarily expressed as a pH value, which is a measure of the relative concentrations of hydrogen (H^+) and hydroxyl (OH^-) ions in solution. The pH scale ranges from 0–14, with pH 7 being the neutral point where the number of H^+ ions balance the OH^- ions. When there are more H^+ ions than OH^- ions the material is acid and the pH will be below 7. Conversely, when the OH^- ions exceed the H^+ ions the material will be alkaline with a pH value above 7. Technically speaking, the pH is the common logarithm of the reciprocal of the H^+ ion concentration, and as the pH values are in a logarithmic scale this means that pH 5 is ten times as acid as pH 6, and pH 4 is 100 times as acid as pH 6.

4.6.1   pH measurement

This can be made either with indicator solutions, whose change in colour is related to the H^+ ion concentration, or potentiometrically with a pH meter and a glass electrode system. There are a number of indicators that can be used to measure the pH. Bromo-cresol green changes from yellow at pH 3.6 to blue at pH 5.2, chlorophenol red changes from yellow at pH 4.6 to violet at pH 7.0, and bromo-thymol blue changes from yellow at pH 6.0 to blue at pH 7.6. To overcome the relatively narrow bands over which these indicators work, a universal indicator with a range from pH 3.0 to pH 11.0 is available, but there is some loss of accuracy as readings cannot be estimated closer than 0.5 of a pH unit. A wide-range indicator can be prepared from a mixture of two parts of bromo-thymol blue and one part of methyl red. The range of colour changes in relation to the soil pH are:

Colour	pH
brilliant red	<3.0
red	3.1–4.0
red-orange	4.1–4.7
orange	4.8–5.2
orange-yellow	5.3–5.7
yellow	5.8–6.1
greenish-yellow	6.2–6.4
yellowish-green	6.5–6.7
green	6.8–7.3
greenish-blue	7.4–7.8
blue	>7.9

However, when this indicator is used on some peats a preferential adsorption of one of the dyes can cause misleading results. Indicator solutions are useful for quick tests, but they cannot give the accuracy of a properly conducted test using a pH meter.

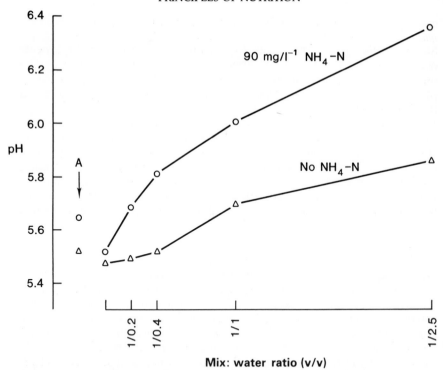

Figure 4.8 The effects of medium:water ratios and ammonium nitrogen on the measured pH of the medium (redrawn from Bunt & Adams 1966a).
Key: A – determinations made with suspensions in 0.01 M CaCl$_2$ (1:2.5 ratio).

Erroneous readings can, however, be obtained with a pH meter if the test is not made correctly. The conventional method of making a pH test on a suspension of medium and distilled water may give pH readings that are significantly higher than those experienced by the plants growing in the medium. An example of the way in which the medium:water ratio can affect the result of pH determinations is shown in Figure 4.8. Increasing the ratio from a saturated paste to 1:2.5 (mix:water) by the addition of water raised the pH by 0.35 units if the mix did not contain ammonium nitrogen, and by 0.85 units if 90 mg l^{-1} ammonium nitrogen was in the mix. If a weak electrolyte solution (e.g. 0.01 M calcium chloride solution) was used instead of water the pH was very similar to the saturated paste determinations, i.e. using a weak electrolyte solution instead of distilled water largely overcame the dilution effect.

4.6.2 Minor element availability

The pH value of the mix is important for a number of reasons. Although at very low pH values the H$^+$ ions can themselves cause toxicity, it has been shown that plants are able to grow over a wide range of H$^+$ ion concentration, from pH 4 to pH 8, before the acidity or alkalinity as such causes any trouble, *provided that the minor elements required by the plant are maintained in an available form* (Arnon & Johnson 1942). Usually the pH will be of concern to the grower and the soil chemist because of its effects on the availability of plant nutrients, not all of which behave in the same manner. Lucas & Davis (1961) have drawn attention to the way in which pH affects the availabilities of major and minor nutrients in organic soils, they concluded that the optimal pH range in these soils was 5.0–5.5, i.e. 1.0–1.5 units lower than for mineral soils. The strong contrast between nutrient availability in mineral soils and a lightweight potting mix is shown in Figure 4.9. The potting mix consisted of sphagnum peat, vermiculite, perlite, composted bark and sand (Metro-Mix 300[R]), and Figure 4.9b shows the sharply decreasing availabilities of phosphorus, iron, manganese and boron in this mix as the pH increases. These results confirm the desirability of maintaining the pH of organic and lightweight mixes in the 5.0–5.5 region.

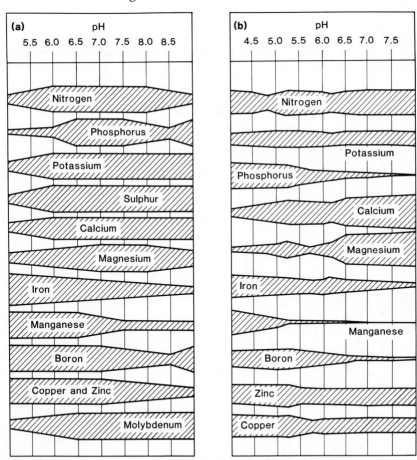

Figure 4.9 Contrasting effects of pH on nutrient availabilities in (a) mineral soils (after Truog 1948) and (b) a typical lightweight potting mix (after Peterson 1981).

4.7 LIME REQUIREMENT

Although a pH determination measures the *active* soil acidity it does not give a direct indication of the *reserve*, or *exchange*, acidity, i.e. it is an intensity factor like temperature. It is possible for two materials to have the same pH, or active acidity, but significantly different amounts of reserve acidity. They will, therefore, have different lime requirements. Mixes containing large quantities of sand, perlite and expanded plastics, for example, will require only relatively small quantities of lime, whereas mixes made from peats and mineral soils having a high clay content will require much more lime to effect a given increase in the pH. Such materials are said to have a high lime requirement and to be well buffered against a change in pH.

The lime requirement of four mixes made from 75% by volume of peat and 25% of sand is shown in Figure 4.10; three of the peats were of the sphagnum type, originating from different countries, and the fourth was a sedge peat. The pH of the sedge peat mix, before any lime had been added, was about 1 pH unit higher than any of the sphagnum peat mixes, but its rise in pH with increasing amounts of lime was relatively low. Although the initial pH of the sphagnum peat mixes, before any lime had

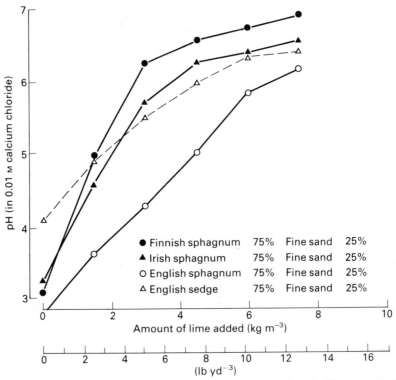

Figure 4.10 The lime requirement of four peat–sand mixes made with different peats. The liming material was composed of equal parts of calcium carbonate and Dolomite limestone.

been added, was very similar for all peat sources, the Finnish peat mix had a slightly lower lime requirement than the Irish peat mix, and the English sphagnum peat mix had a very high lime requirement, almost twice as much lime being required to reach pH 6 as for the other two sphagnum mixes. For practical purposes, the lime requirements over the linear part of the curves for mixes made from these different peats are:

Medium	Lime required to raise the pH by 1 unit
75% Finnish sphagnum, 25% fine sand	$0.9 \, kg \, m^{-3}$ ($1 \, lb \, 8 \, oz \, yd^{-3}$)
75% Irish sphagnum, 25% fine sand	$1.1 \, kg \, m^{-3}$ ($1 \, lb \, 14 \, oz \, yd^{-3}$)
75% English sphagnum, 25% fine sand	$2 \, kg \, m^{-3}$ ($3 \, lb \, 6 \, oz \, yd^{-3}$)
75% English sedge, 25% fine sand	$2 \, kg \, m^{-3}$ ($3 \, lb \, 6 \, oz \, yd^{-3}$)

These results show the need to know the exact lime requirement of each medium rather than work to rule-of-thumb measurements with unknown materials.

In this experiment the 'lime' consisted of equal quantities of calcium carbonate and Dolomite limestone. To avoid any extraneous effects, other fertilizers were not included. The normal base fertilizers used in the GCRI peat–sand mix (§ 8.8) would have reduced the pH by approximately 0.5 units. This is mainly due to the acidifying action of the superphosphate (see Table 6.1); other fertilizer mixtures could be expected to give slightly different results. All pH determinations were made after an equilibrium between the lime and the moist medium had been reached.

4.7.1   Increasing the pH of media in containers

Often the pH of media in containers will rise with time because of bicarbonates in the irrigation water, or the use of alkaline feeds; ways of countering this effect are discussed in section 4.7.2. However, sometimes the pH is found to be too low, either because of feeding with acidic fertilizers or due to failure to add sufficient lime during the mixing. Where the pH is falling slowly it can be corrected by liquid feeding with calcium nitrate or by applying limewater solutions (made by adding 1 kg calcium hydroxide to 1000 l of water (1 lb to 100 gal), allowing to stand overnight and applying the clear solution). The increase in the pH of the medium will, however, be small; calcium hydroxide has a low solubility and most media have fairly high lime requirements.

Two other methods have been used to quickly raise the pH of mixes while in the containers. Jarrell et al. (1979) applied solutions of either potassium or sodium bicarbonates to plants growing in mixes with a pH of 4.0. There was an immediate rise in the pH, followed by a steady decline during the next 40 days. The required strength of the bicarbonate solution will depend upon the pH of the mix and its buffer capacity; their results suggest that solutions of 0.05–0.1 M of either potassium bicarbonate or sodium bicarbonate should be sufficient. There was no reported injury to the plants. However, it must be remembered that although this treatment raises the pH, no calcium is being supplied so problems might occur with calcium deficiency even at high pH values. Keisling & Lipe (1981) compared several ways of raising the pH of a 3:1:1 peat:perlite:vermiculite mix in which poinsettias were growing (pH 4.7–5.3). Applying limestone or calcium hydroxide to the surface of the medium at rates equivalent to 1.2 kg m^{-3} calcium carbonate only raised the pH of the surface layer, below the 2.5 cm depth there was no effect. Punching 6 mm diameter holes in the medium before liming raised the pH in the top 5 cm; and injection of a limestone slurry raised the pH at all depths. It is obviously much easier to raise the pH of the medium before potting than afterwards.

4.7.2 Decreasing the pH of the media

Most of the materials used to make potting mixes are either strongly acid, e.g. sphagnum peat, or neutral to slightly alkaline, e.g. some sources of vermiculite. In the latter case, adding peat to the vermiculite will usually decrease the pH of the mix sufficiently to avoid nutritional problems. With some materials that have a high pH, and would therefore be unacceptable for use in potting mixes, it is possible to lower the pH by treatment with sulphur; this is oxidized in the media to sulphuric acid by *Thiobacillus* species:

$$S + 3/2\ O_2 + H_2O \rightarrow H_2SO_4$$

The equivalent weights of sulphuric acid and calcium carbonate or chalk are almost equal, so one unit of acid will react with one unit of chalk in the mix.

Just as the lime requirements of peats differ, so do the amounts of sulphur required to achieve a given pH reduction. The peats in Figure 4.11 had been used to grow a crop of tomatoes and the pH had risen because of bicarbonates in the irrigation water. Sphagnum D had the lowest buffer capacity and required only 530 g m^{-3} of sulphur to reduce the pH to 5.5; to obtain the same pH sphagnum C required 875 g m^{-3}, and the sedge peat required 1.5 kg m^{-3}, i.e. three times as much sulphur as sphagnum D. Spent mushroom compost had a higher buffer action than any of the peats, 2.5 kg m^{-3} of sulphur only reduced the pH to 6.0. Working with a mixture of sewage sludge, peat and sand, Gouin & Link (1982) found that 10 kg m^{-3} of sulphur (equal parts of wettable and granular sulphur) reduced the pH from 6.9 to 4.4. For rapid results the sulphur particles must be fine, i.e. they must pass through an 80-mesh sieve, the medium must be moist but warm and well aerated. Most of the reaction should then occur within 4–6 weeks.

Several sulphur compounds can also be used to acidify the growing media, some of those commonly used and their sulphur equivalents

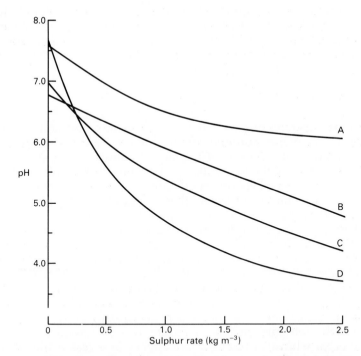

Figure 4.11 Changes in media pH with rates of sulphur application.
Key: A spent mushroom compost; B old tomato modules, sedge peat; C old tomato modules, sphagnum peat, medium decomposition; D old tomato modules, young sphagnum peat.

are given in Table 4.10. Iron and aluminium sulphates can either be added during mixing or as a drench to the growing plants, usually at the rate of 3 g l^{-1} (0.5 oz gal^{-1}). As well as lowering the pH by the formation of sulphuric acid:

$$Al_2(SO_4)_3 + 6H_2O \rightarrow 2Al(OH)_3 + 3H_2SO_4$$

the aluminium replaces any exchangeable hydrogen and this increases the acidification effect further. A similar effect is obtained with ferrous

Table 4.10 Sulphur-containing materials that can be used for acidification.

Material	Formula	Sulphur content (%)
aluminium sulphate	$Al_2SO_4 \cdot 18H_2O$	14.4
ammonium sulphate	$(NH_4)_2SO_4$	24.2
ferrous ammonium sulphate	$Fe(NH_4)_2(SO_4)_2$	16.0
ferrous sulphate	$FeSO_4 \cdot 7H_2O$	11.5
sulphuric acid (93%)	H_2SO_4	30.4
sulphur	S	100

sulphate. However, some caution is required in using media treated with high rates of acidifying chemicals, as plants have sometimes shown foliar necrosis. Nitrification rates in soils acidified with sulphur can remain depressed even when the soils have been limed to raise the pH again.

4.8   SOLUBLE SALTS (OSMOTIC POTENTIAL)

To achieve the high rates of growth required in modern pot plant culture, the levels of the major nutrients in the medium are maintained at relatively high concentrations by using large amounts of base fertilizers and high strengths of liquid fertilizers. To some extent this practice compensates for the small volume of medium available to the plant roots and the rapid depletion of nutrients that would otherwise follow. We have already seen that the nutrients required by plants are present in varying degrees in the soil solution, and although an increase in their concentration will often result in an increased growth rate, very high concentrations can restrict growth or even cause the death of the plant. This build-up in nutrients, which is often referred to as a high salinity, operates either as a specific ion toxicity or as a general salinity effect by reducing the availability of water to the plants. Specific ion toxicities, such as those of manganese or boron, occur only infrequently, whereas a high level of soluble nutrients, mainly of nitrate and potassium in potting mixes, can occur whenever fertilizers are applied in excess of the rate of plant uptake and the loss by leaching.

4.8.1   Plant response

Water stress within the plant is one of the factors which directly control plant growth. The magnitude of the stress is controlled by:

(a) the water stress within the aerial environment, i.e. the vapour pressure deficit, which affects the transpiration rate;
(b) the water stress within the medium which affects the rate of water absorption by the plant.

It is the latter with which we are now concerned.

We have already seen that the total soil moisture stress comprises the matric tension and the solute stress. Both of these components affect plant growth by reducing the availability of the soil water and may thereby produce a water deficit within the plant. A number of workers have demonstrated that salts dissolved in the soil solution can reduce the availability of water to plants. Eaton (1941) used the 'split-root' technique to show that when the roots of corn plants were equally divided between two nutrient solutions, one with a salt stress of approximately $\frac{1}{3}$ of an atmosphere (33 kPa) and the other $1\frac{3}{4}$ atmospheres (175 kPa), the plants took up water from the weak solution at approximately twice the rate at which water was absorbed from the stronger solution. It can also be shown that, under certain conditions, the effects of matric tension and solute stress in soils are additive in their effects on plant growth. Wadleigh & Ayers (1945) grew bean plants in soil cultures which were allowed to dry out to different soil water tensions before they were watered. Within each set of water tension treatments, a range of osmotic stress treatments was obtained by adding different amounts of sodium chloride to the soils. Their results, reproduced in Figure 4.12, show that the plants integrated the two separate factors, and responded to the total soil moisture stress irrespective of its composition. Subsequent work has shown, however, that this concept may be an oversimplification. Plants are able to adjust to an increase in the soil osmotic potential by making a corresponding increase in the osmotic potential of their cells. Although a sudden increase in the salinity of the soil can cause the plants to wilt, they are usually able to adjust to the higher soil salinity at a rate of about one atmosphere (100 kPa) per day (Bernstein 1963). Turgidity will then be regained, unless the roots have received permanent damage. Another possible cause of the reduction in plant growth, observed under saline conditions, is the specific ion toxicity effect. For example, Gauch & Wadleigh (1944) found, when growing beans in water cultures of varying solute stress induced by different salts, that growth was inhibited more by

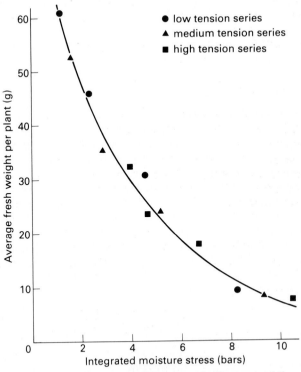

Figure 4.12 Plant growth responds to both water and salt stress and these components have been integrated into the total soil moisture stress (from Wadleigh & Ayers 1945).

magnesium sulphate and magnesium chloride solutions than it was by sodium sulphate and sodium chloride solutions of a similar solute stress. Thus the growth of plants in soil with a high solute stress will be reduced because soil water is less available, and growth may be further reduced by a specific ion toxicity. The response of the plant to soil salinity will also be determined by the aerial environment in which it is growing. When the vapour pressure deficit is high, as occurs in a glasshouse in summer under conditions of high solar radiation, high temperature and low humidity, the reduction in growth due to a high soil moisture stress will be much greater than in winter, when the vapour pressure deficit is less and the rate of transpirational loss is lower.

4.8.2 Salinity control in containers

The main reasons why the salinity of a mix can increase after being in a container are:

(a) insoluble fertilizers, such as hoof and horn and the controlled-release fertilizers 'Osmocote' or 'Nutricote', either become mineralized to produce nitrates or release salts by diffusion in excess of the amounts absorbed or leached;
(b) the quantity of salts supplied by the irrigation water or liquid feed is greater than the amounts absorbed by the plant and lost by leaching.

Such situations can largely be prevented by appreciating the amounts of fertilizers required by crops under different environments and avoiding excessive applications. A rise in the salinity can be prevented, or corrected if it has already occurred, by controlled leaching.

Prevention of salinity increase   Normally the solutes in the soil solution will be at their highest concentration just before the plant is irrigated. The soil solution will be displaced down the container by the irrigation water or liquid feed, and if enough of this low salinity solution is applied some of the higher salinity soil solution will be lost by leaching. Normally a leaching fraction of 10–30% will be sufficient to prevent the salinity of the mix from increasing. For plants such as *Azalea*, which require a low salinity, Kofranek & Lunt (1975) recommend that the root zone concentration does not exceed 15 meq l^{-1}; the electrical conductivity (EC) of the soil solution would then be equivalent to about 1.5 mS cm^{-1} and a saturated media extract would be approximately 0.75 mS cm^{-1}. To maintain this level the calculated leaching requirement (LR) using water with an EC of 0.5 mS cm^{-1} will be:

$$LR = \frac{EC \text{ irrigation water}}{EC \text{ drainage water}}$$

$$= \frac{0.5 \text{ mS cm}^{-1}}{1.5 \text{ mS cm}^{-1}}$$

$$= 0.33 \text{ or } 33\%$$

That is, 33% of the total irrigation water with an EC of 0.5 mS cm^{-1} would have to be discharged as drainage to maintain an EC in the root zone of 1.5 mS cm^{-1}. If a more saline solution is used for irrigation then the leaching requirement will be higher, e.g.:

$$\frac{1.0 \text{ mS cm}^{-1} \text{ irrigation water}}{1.5 \text{ mS cm}^{-1} \text{ drainage water}} = 0.66 \text{ or } 66\%$$

Leaching of high salinity mixes   If the salinity of the mix is already high it will be necessary to use more water, and the efficiency with which the salts can be displaced depends upon certain physical characteristics of the mix. Rose (1973) discussed the leaching of salts with respect to several parameters, including particle size, particle porosity, diffusion of solutes from within the aggregate and the velocity of the leaching solution within the medium. The most important factors controlling leaching efficiency were the size of the particles or aggregates and the flow velocity. Maximum efficiency occurs only within a narrow band of flow velocity ×️ aggregate size, at values above and below this band the efficiency decreases sharply. In media with large connecting pores channelling occurs, this is not as efficient in removing salts as a uniform or 'piston flow' which occurs in fine, closely packed media. The results obtained by Kerr (1983) with a range of potting mixes illustrate this point. The efficiency of salt removal decreased as the permeability of the mix increased. Most of the salt was removed by applying 1.5 times the amount of water held in the mix at container capacity.

Where high salinity of the mix is suspected the leachate should be monitored. If confirmed, leaching with good quality water to give the equivalent of one container volume of leachate should correct the problem. General aspects of the quality of irrigation water are discussed in section 9.5. Other measures to mitigate the effects of salinity are:

(a) keep the medium wet,
(b) never apply either powdered fertilizer or a strong liquid feed when the medium is dry,
(c) reduce the stress on plants by shading and raising the humidity.

The extent to which salinity affects plants depends upon their age (see Fig. 8.2), local environment, management practices and also the species. Examples of plants having different degrees of tolerance are:

(a) **very sensitive**: *Azalea, Camellia, Cytisus* x *praecox* 'Moonlight', *Gardenia, Pittosporum tobira* 'Variegata', *Primula, Mahonia aquifolium* 'Compacta';
(b) **sensitive**: *Aphelandra, Clivia miniata, Erica, Ficus benjamina*, lettuce, many bedding plants;
(c) **tolerant**: carnation, *Chrysanthemum, Cupressus arizonica, Diffenbachia, Hydrangea, Magnolia grandifolia, Philodendron*, tomato;
(d) **very tolerant**: *Acacia cyanophylla, Atriplex, Bougainvillea* 'Barbara Karst', *Callistemon citrinus, Cordyline indivisa, Dietes vegeta, Hibiscus rosa-sinensis, Spartium junceum, Yucca aloifolia*.

4.8.3 Salinity measurement

It is important that the level of soluble salts in growing media is monitored frequently, this is usually done by measuring the electrical conductivity of either a saturated extract or of a suspension in water. The operating principle is that the greater the concentration of fertilizers or salts, the greater will be the conductance, which indicates the osmotic concentration of the soil solution. This method however only measures the electrolytes that are in solution; if urea or other organic compounds that do not ionize are present in significant amounts then the osmotic concentration will be greater than that indicated by the conductivity reading (see Fig. 9.2 for the effect of urea concentration on conductivity).

The unit used to measure conductance has been the mho, which is the reciprocal of the unit of resistance, i.e. the ohm. The SI unit which is gradually replacing the mho is the Siemen. However, both units, which have the same value, are large in relation to the salt concentration found in growing media. To avoid using large numbers of decimal places before the conductivity number, a smaller unit of conductivity is used. The units commonly used are:

Units	Value	Ways of writing the same salt concentration	Notes
μmho cm^{-1} (micromho) μSiemen (μS) cm^{-1} (microSiemen)	mho Siemen $\Big\} \times 10^{-6}$	$2000\,\mu$S	These units are used in the UK by ADAS
mho cm$^{-1} \times 10^{-5}$ SC (specific conductivity)	mho Siemen $\Big\} \times 10^{-5}$	200×10^{-5} S	In the USA, older models of Solu Bridge are calibrated to read from 10–1000 mho $\times$ 10^{-5}. In Ireland, the salinity is reported as the specific conductivity (SC), which equals mho $\times 10^{-5}$.
CF (conductivity factor)	mho Siemen $\Big\} \times 10^{-4}$	CF 20	Many of the conductivity meters used by growers in the UK are calibrated to read from 0.1 to 100 mho $\times 10^{-4}$
mmho cm^{-1} (millimho) mSiemen cm^{-1} (milliSiemen)	mho Siemen $\Big\} \times 10^{-3}$	2 mmho	New models of Solu Bridge meters are calibrated to read from 0.1 to 10 mho $\times 10^{-3}$.

Not only is it necessary to know which of the above units are being used, it is equally important to know how much water has been added to the medium to make the conductivity determination. The following are the methods most commonly used:

Saturated media extract (EC_e) This method was first used for field soils and the term 'saturated paste', which is more descriptive of prepared mineral soils rather than some of the present-day lightweight potting mixes, is still used by some laboratories. Distilled water is added to the sample, while gently stirring, until the medium is saturated. It is left for two hours to equilibrate and is then checked; there should be no free water on the surface. The water and dissolved salts are withdrawn from the medium with a vacuum or water pump, and the electrical conductivity measured at a temperature of 25 °C.

This method of salinity determination has the great advantage that the amount of water used is directly related to the amount of water that can be held by the medium in the container, hence the salinity can be related to that experienced by the plant. The approximate salt stress or osmotic concentration in the medium can be obtained from:

$$\text{osmotic concentration (atm)} = 0.36 \times EC_e \, (\text{mmho cm}^{-1} \text{ at 25 °C})$$

The factor of 0.36 was derived from numerous determinations of the EC and the osmotic pressure of various salt solutions by the USDA Salinity Laboratory (Richards 1954).

The EC can also be used to obtain the approximate salt content of the medium:

$$\text{p.p.m. salts} = EC_e \times 700 \times \text{moisture factor of the medium}$$

The factor of 700 is an average value for converting mS cm^{-1} into p.p.m. of salts in water. The moisture factor is the ratio of the percentage water at saturation (dry weight basis) to percentage water in the mix at container capacity:

$$\text{moisture factor} = \frac{\% \text{ water by weight at saturation}}{\% \text{ water by weight at container capacity}}$$

The two major components which determine the moisture factor are (a) the physical properties of the medium with respect to its water retention, i.e. the relative amounts of organic material and minerals present, and (b) the depth of the container (§ 3.3.2). Waters *et al.* (1972) have reported moisture factors ranging from 1.14 for peat in shallow (5.7 cm, $2\frac{1}{4}$ in) containers to 1.8 for builders sand in 12.7 cm (5 in) pots. A value of 1.4 could be assumed for most situations.

1 : 2 volume procedure An alternative way of determining the salinity of potting mixes is by making a suspension. Although there is no universal agreement on the ratio of growing media to distilled water for making

Table 4.11 Interpretation of salinity readings determined by the saturated media extract and 1:2 volume methods.

Saturated media extract ($mS\,cm^{-1}$ or $mmho\,cm^{-1}$)	1:2 volume ($mSiemen\,cm^{-1}$ or $mmho\,cm^{-1}$)	
<0.74	<0.15	very low
0.75–1.99	0.15–0.50	suitable for seedlings and media high in organic matter, too low if media low in organic matter
2.0–3.49	0.50–1.80	satisfactory for most plants, growth of some sensitive plants reduced
3.50–5.00	1.80–2.25	slightly high for most plants, only suitable for vigorous plants
5.0–6.0	2.25–3.40	reduced growth, plants stunted, wilting and marginal leaf burn
>6.0	>3.40	severe injury and probable crop failure

suspensions, the most common is one part medium by volume to two volumes of water. Volume measurement of medium is preferable to weight as this partially overcomes the large variations in bulk densities of the mixes. It is also desirable that the mix should first be air-dried otherwise variable amounts of water in the samples lowers the accuracy of the determination.

Results can be presented either as $mS\,cm^{-1}$ or as p.p.m. of salt. The latter is calculated from:

$$\text{p.p.m. salt} = mS\,cm^{-1} \times 700 \times 2$$

Guidelines for salinity values based on the saturated media extract and the 1:2 volume method are given in Table 4.11, and for the Dutch 1:1.5 and English 1:6 (ADAS) methods in Table 4.12.

Table 4.12 Interpretation of salinity readings by the Dutch 1:1.5 volume extract and the English 1:6 volume extract.

Method	Low	Fairly low	Moderate	Fairly high	High	Very high
Dutch ($mS\,cm^{-1}$ or $mmho\,cm^{-1}$)	<0.7	0.7–1.2	1.3–1.8	1.9–2.7	2.8–3.6	>3.6
English ($\mu S\,cm^{-1}$ or $\mu mho\,cm^{-1}$)	<150	151–300	301–500	501–700	701–900	>900

1:5 volume procedure   Some laboratories use a wider ratio of medium to water, the salinity values will then be approximately 40% of those obtained by the 1:2 method, unless the medium contains much sulphate.

4.8.4   Salinity measurement of loam-based mixes

The salinity of loam-based mixes can be measured using the saturated media extraction method or by making a suspension. If the loam contains a large amount of calcium sulphate, however, then determinations made with a wide ratio of water:medium will result in more calcium sulphate being dissolved than occurs in the soil solution, thereby giving an erroneously high salinity reading. With such soils this effect can be overcome by making a suspension using a saturated solution of calcium sulphate instead of distilled water (Winsor *et al.* 1963). Although this automatically increases the salinity reading, it also overcomes the effect of variable amounts of calcium sulphate being dissolved in the suspension. Calcium sulphate has a solubility of 0.24 g 100 cm^{-3} with a specific conductivity of 1960 μS cm^{-1} at 20 °C. For determinations based on 1 volume of medium to 2.5 volumes of saturated calcium sulphate, values below 2700 μS cm^{-1} are considered safe, with more than 3000 μS cm^{-1} being dangerous. This method is more applicable to loam-based than to loamless potting media.

CHAPTER FIVE

Nitrogen

From the viewpoint of the plant physiologist, all of the *essential* elements can be regarded as being of equal importance in the nutrition of the plant, but in terms of the practical management of lightweight mixes nitrogen is probably the most important single element. It forms between 2 and 4% of the dry weight of the average pot plant and occurs in all plant proteins. It is also an essential component of chlorophyll, the green pigment present in leaves which enables plants to utilise the sun's energy to combine carbon dioxide from the atmosphere with water vapour to form carbohydrates. This process is known as photosynthesis and provides man with his primary source of food. When nitrogen is deficient there is a marked reduction in the growth rate. Plants are tall and spindly, the upper leaves being narrow and erect while the lower leaves turn first pale green and then yellow as the nitrogen is withdrawn and transported to the actively growing regions. Often the older leaves produce bright colours before they die. When the nitrogen level in the medium is high in relation to the amounts of phosphorus and potassium, plants tend to make soft, vegetative growth and be less reproductive with delayed flowering. This situation is intensified when the plants are grown under conditions of low light intensity, high humidity and low water stress. Soft vegetative plants are more susceptible to fungal diseases such as botrytis. If the nitrogen level in the medium is further increased there will be a corresponding increase in the soluble salt level, and eventually this will have the effect of producing hard, stunted plants.

Although the atmosphere contains approximately 80% of gaseous nitrogen, this is not available to the plants, apart from those cases where there is a nitrogen-fixing symbiotic association with certain soil bacteria. Excluding the few cases where nitrogen compounds are applied to plants as foliar sprays, notably the pineapple, virtually all of the nitrogen required must be absorbed through the roots in an inorganic form, predominantly as ammonium (NH_4^+) or nitrate (NO_3^-).

5.1 NITROGEN AND POT PLANTS

There are three main reasons why a continuous supply of mineral nitrogen is of relatively greater importance when growing pot plants in loamless mixes than it is for border and field grown crops.

Firstly, many of the materials used to make lightweight potting mixes are not able to supply significant amounts of nitrogen to the plants. Some, such as perlite and vermiculite, do not have any mineral nitrogen, others may actually reduce temporarily the available nitrogen by either chemical adsorption, e.g. pine bark, or by biological fixation, e.g. some sphagnum peats. Other materials are able to supply varying amounts of nitrogen, e.g. humidified sphagnum or sedge peats and spent mushroom compost. Peats contain 1–3% by weight of nitrogen, but the actual amount available to the plants varies with the peat type. Comparison of the amounts of available mineral nitrogen in three peats, measured by the amount of nitrogen absorbed by tomato plants grown in 750 cm^3 containers without losses by leaching, are shown in Figure 5.1. Other nutrients had been added to all media and were not limiting. The loam compost was made with 58% by volume of an old arable loam, it would be considered as being too low in organic matter (and available nitrogen) to make a good John Innes loam-based potting compost. Plants grown in the sedge peat had 15% more nitrogen than those in the loam compost, whereas those in sphagnum A, a young peat with little humification, had only 46% as much nitrogen as those in the loam compost. The amount of available

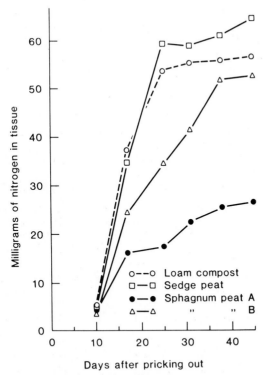

Figure 5.1 Rates of nitrogen absorption by tomato plants grown in a mineral soil compost or in various peats. With the exception of nitrogen, all nutrients had been added in sufficient amounts not to become limiting factors.

nitrogen in most soil-less media is therefore small in comparison with the amounts present in well grown container and bedding plants (Table 5.1 and Fig. 5.2).

Second, the roots of plants in containers are restricted to a small volume; unlike crops in fields and borders, plants cannot rely on an expanding root system to provide a continuing supply of nitrogen. Third, most lightweight mixes are not able to adsorb and retain significant amounts of mineral nitrogen against leaching, ammonium and nitrate nitrogen are readily lost by drainage.

The total amount of nitrogen and the rate at which it is required by the plant depend largely upon the vigour of the species, and the manner in which the plants are grown. Temperature, light, water availability and size of container are some of the factors which control the rate and amount of growth. Some typical plant weights with the amount of nitrogen in their tissue, and the period over which they were grown, are given in Table 5.1. The contrast between the rates at which nitrogen was absorbed by pot chrysanthemums grown in summer, a crop which has a very high nitrogen requirement, and by the relatively slower growing cyclamen, is shown in Figure 5.2; both crops were grown in containers holding one litre of mix. During the last eight weeks of its ten-week growing period, the pot chrysanthemum absorbed nitrogen at the rate of 30 mg day^{-1}, the total nitrogen uptake being 1870 mg. With the cyclamen, the maximum rate of nitrogen absorption between August and December was only 7 mg day^{-1}, and the total nitrogen uptake over the one-year growing period was 1030 mg. In its ten-week growing period the chrysanthemum had assimilated almost twice as much nitrogen as the cyclamen had during a whole year.

Table 5.1 The range of dry weights and nitrogen contents of some pot and bedding plants.

Plant	Total dry weight (g)	% nitrogen	Total nitrogen (mg)	Growing period
Saintpaulia	3.2	2.1	67	15 weeks
Exacum	3.6	4.0	144	16 weeks
Tomato	4.1	3.7	152	10 weeks (Nov.–Jan. propagated)
box of bedding begonias (32 plants/ 4.5 l^{-1} of media)	16.2	3.7	599	8 weeks (summer)
Zinnia (ditto)	25.4	1.4	355	8 weeks (summer)
Chrysanthemum				
winter	20.0	5.5	1030	10 weeks
summer	57.0	3.3	1870	10 weeks
Cyclamen	26.6	3.9	1037	1 year
Ficus	66.4	1.6	1062	1 year

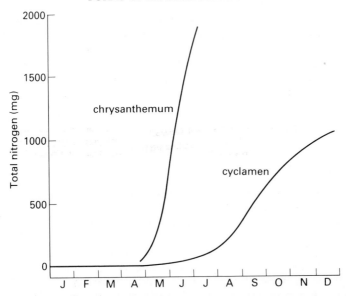

Figure 5.2 Contrasting the rates and the total amount of nitrogen taken up by a pot chrysanthemum, grown for ten weeks in summer and a cyclamen, grown over one year. Both plants were grown in one litre of mix and were given liquid feeding.

The maximum amount of mineral nitrogen that can safely be added in the base fertilizer varies with the plant species, but it is only of the order of 200–250 mg l^{-1} of mix. Quantities in excess of this can cause either a specific toxicity or a high salinity hazard. Nitrate nitrogen causes a big increase in the mix salinity, and although ammonium produced from urea does not ionize to any appreciable extent, and so is not detected by salinity measurements based on electrical conductivity, this source of nitrogen does, in fact, contribute to the osmotic potential or the total salt stress experienced by the plant. Therefore only a small part of the total amount of nitrogen required by the average pot-grown plant can be supplied as mineral nitrogen in the base fertilizer. Either the greater part of the nitrogen must be supplied in a form which gives a controlled release over the growing period, or the relatively small amount of mineral nitrogen that can safely be applied in the base fertilizer must be supplemented by frequent liquid feeding. This latter system of nutrition is discussed in detail in Chapter 9.

5.2 FORMS OF MINERAL NITROGEN

As nitrogen is required by plants in large amounts, and can be absorbed either as cations (NH_4^+) or anions (NO_3^-) the form of nitrogen can have a large effect on the ionic balance within the plant. Normally the concentration of the inorganic cations within the plant (K^+, Na^+, Ca^{2+}, Mg^{2+}) exceeds the concentration of the inorganic anions (NO_3^-, $H_2PO_4^-$, SO_4^{2-},

Cl^-), the balance between them consisting of organic acid anions, usually malate, oxalate and critrate. This is sometimes referred to as the (C–A) balance. Plants supplied with NH_4^+ have lower concentrations of inorganic cations, Ca^{2+}, Mg^{2+} and K^+, but higher concentrations of inorganic anions $H_2PO_4^-$, SO_4^{2-} and Cl^- than plants receiving NO_3^- nitrogen. Plants absorbing NH_4^+ also have lower concentrations of organic acids.

5.2.1   Ammonium and nitrate nitrogen

Plants absorb only relatively small quantities of the various organic forms of nitrogen and most of the mineral nitrogen is taken up in the ammonium and nitrate forms. Frequently, no clear preference is shown by plants for either of these forms of nitrogen, but there are certain exceptions, e.g. the inhibition of flowering in Lemna perpusilla by low concentrations of the ammonium ion but not by nitrate ions (Hillman & Posner 1971).

The effects which ammonium and nitrate forms of nitrogen can have on plant growth have received wide attention; however, most of the experiments have been made either in sand and water cultures or with field crops grown in mineral soils. The response to ammonium and nitrate nitrogen is dependent upon a number of factors, the most important are: the species of plant, the ratio of ammonium nitrogen to nitrate nitrogen in the growing medium and its pH, the relative availability of other cations, anions and some of the microelements, and the plant environment. In general, high concentrations of ammonium when nitrates are low or absent can cause the following detrimental effects:

(a) a reduction in the rates of photosynthesis and plant growth;
(b) epinastic curling of leaves, chlorosis and stem lesions;
(c) roots darkened, restricted and possibly killed;
(d) reduced ability of roots to absorb water, leading to loss of turgor at times;
(e) reduced plant uptake of other cations, i.e. calcium, magnesium and potassium;
(f) an increase in the pH of the plant tissue and lower (C–A) values.

Among the beneficial effects of including a moderate amount of ammonium nitrogen in preference to using only nitrate nitrogen are: earlier maturity of some crops and often a better leaf colour.

Ammonium is readily absorbed by plant roots and, unlike nitrates, does not have to be reduced before being incorporated into organic compounds. Whereas plants are able to accumulate and transport nitrate nitrogen without major toxic effects, the accumulation of ammonium nitrogen in the leaves is usually highly toxic. Ammonium and carbohydrates are used to synthesize organic molecules, and high rates of ammonium absorption can lead to reduced photosynthetic activity and to carbohydrate depletion. Because seedlings and small plants growing under low light conditions in winter have low carbohydrate supplies, they are more liable to suffer from ammonium toxicity than older plants.

The form in which the nitrogen is supplied to the plants can also cause large differences in the media pH. When plants are grown with ammonium as the nitrogen source hydrogen ions are secreted by the plant, leading to a lowering of the media pH:

$$NH_4^+ \rightarrow NH_3 + H^+$$

Not only is there a general reduction in the pH of the medium, the pH of the rhizosphere around the roots can be a whole pH unit lower than the bulk of the medium (Smiley 1974). When nitrates are absorbed by the plant hydroxyl ions are secreted, causing the pH of the medium to rise:

$$NO_3^- + 8H^+ \rightarrow NH_3 + 2H_2O + OH^-$$

Usually the toxicities found when plants are grown with high ammonium levels in the substrate can be reduced or eliminated if the pH of the medium is maintained near to neutrality by adding calcium carbonate. This restricts the accumulation of ammonium in the leaves and also improves the colour and vigour of the roots. However, high pH values, i.e. 6.5 and over, are not recommended when growing plants in mixes based on organic and lightweight materials. Experience has shown that nutrition and management are easier in the pH range 5.0–6.0, at higher values deficiencies of iron and boron can readily occur.

Another problem associated with the use of ammonium fertilizers is the suppression of the amounts of calcium, magnesium and potassium that plants are able to absorb. Tew Schrock & Goldsberry (1982) examined the effects of liquid feeds with varying $NH_4:NO_3$ ratios on geraniums and petunia seedlings when grown in soil and soil-less mixes. Five liquid feeds with $NH_4:NO_3$ ratios of 1:0, 3:1, 1:1, 1:3 and 0:1 were used, the total amount of nitrogen supplied being the same for all feeds. They found that the ratio of $NH_4:NO_3$ had no effect on the height, fresh and dry weights, or number of vegetative shoots on plants grown in soil; in the soil-less medium, however, growth was adversely affected when the proportion of ammonium nitrogen was above 50% (Fig. 5.3). With plants in the soil-less medium, increasing the proportion of ammonium nitrogen caused a strong and progressive decline in the amount of calcium in the tissue, i.e. falling from 3% to 1%. In the soil mix this effect was much less, the calcium fell only from 4% to 3.2%. It is also known that tomato plants grown with ammonium as the nitrogen source have lower calcium levels and are more susceptible to blossom-end rot of the fruit, a physiological disorder related to the calcium supply, than are plants receiving nitrate nitrogen (Wilcox *et al.* 1973).

The reasons why plants grown in loamless mixes are more susceptible to ammonium toxicity than those in soil mixes include:

(a) larger applications of fertilizers of either organic, slow-release or ammonium forms to loamless mixes to compensate for the naturally lower amounts of nitrogen released by the media;

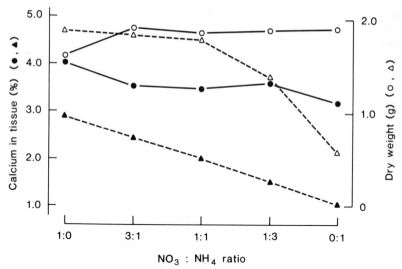

Figure 5.3 Response of petunia to nitrate-N:ammonium-N ratios when grown in soil and soil-less mixes.
Key: soil mix ○ dry weight; ● percentage calcium.
    soil-less mix △ dry weight; ▲ percentage calcium.

(b) the slow rates of nitrification in mixes made with fresh sphagnum peat;
(c) lower pH values than in soil mixes; and
(d) differences in microelement availabilities.

Experience suggests that the $NH_4:NO_3$ ratio should not exceed 1:1; and for some plants grown under low light integrals in winter, 25% ammonium nitrogen with 75% nitrate nitrogen is a better ratio. Plants reported as susceptible to ammonium toxicity include: bean, cucumber, tomato, radish, *Antirrhinum, Chrysanthemum,* geranium, *Impatiens, Petunia, Poinsettia, Saintpaulia* and *Viburnum.* The onion is able to tolerate higher levels of ammonium than most other plants; the bulb acting as an ammonium sink.

Plants that prefer the ammonium form of nitrogen are calcifuge plants in general, and azaleas in particular. Colgrave & Roberts (1956) found that azaleas receiving ammonium nitrogen had better growth and showed less chlorosis than plants supplied with nitrate nitrogen. The chlorosis was related to the pH of the leaf tissue; ammonium nutrition reduced the uptake of other cations, thereby reducing the tissue pH. However, nitrates increased the absorption of cations, raising the pH of the tissue and inactivating the iron in the leaf. Similar results have been observed with blueberries. Other instances of reduced growth and leaf chlorosis when plants have received only nitrate nitrogen have been reported by North & Wallace (1959) for macadamia plants, Conover & Poole (1982a) for *Calathea,* and Nelson & Selby (1974) for Sitka spruce and Scots pine. Conifers grown with nitrate nitrogen showed lower activity of iron in the leaves, this was due to the greater concentration of cations in the tissue and the excess of organic anions. In experiments at the Glasshouse Crops Research Institute with pot chrysanthemums, plants that are not normally associated with iron deficiency problems, the inclusion of some ammonium in the liquid feed has always produced a better leaf colour than an all nitrate feed; however, leaf colour was not always directly related to the pH of the medium.

Apart from the risks of ammonium toxicity, two other hazards which are associated with the use of ammonium-producing fertilizers are free ammonia (NH_3) and nitrite (NO_2^-) toxicities.

Ammonium and free ammonia In mineral soils the ammonium ions, which can be derived from either an ammonium fertilizer, such as ammonium sulphate, or from the mineralization of organic fertilizers, such as hoof and horn and ureaformaldehyde, are rapidly converted into nitrites and then into nitrates. In loamless mixes this process is sometimes restricted if the population of nitrifying bacteria is low, or if these bacteria are temporarily eliminated from the mix following heat pasteurization, thereby causing an accumulation of either ammonium or nitrite nitrogen. A simplified diagram showing the nitrogen transformations which occur in soils is given in Figure 5.4.

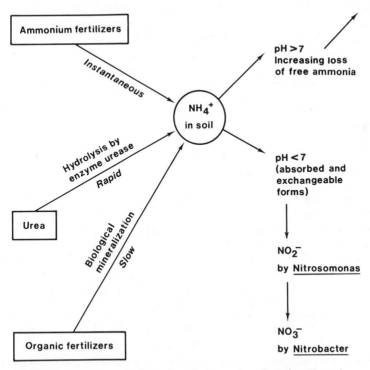

Figure 5.4 Nitrogen transformations that occur in soils and potting mixes.

In soils containing ammonium ions, free ammonia can occur when the pH is close to the neutral point, and its concentration increases as the pH rises. At pH 7 about 1% of the ammonia is non-ionized and at pH 8 this increases to 10%. The rise in pH of a peat–sand mix made with organic nitrogen, and the accompanying production of free ammonia is shown in Figure 5.5. Free ammonia readily penetrates cell membranes and is therefore more toxic to plants than the ammonium ion. The effect of free ammonia on the growth of sudan grass and cotton seedlings was investigated by Bennett & Adams (1970); they concluded that the initial concentration for incipient toxicity was 0.15–0.20 mM of non-ionized ammonia in the soil solution. In experiments at the Glasshouse Crops Research Institute, the maximum recorded loss of free ammonia occurring over a period of 42 days was 19.5 mg nitrogen l^{-1} of medium. A total of 850 mg of organic nitrogen had been applied, and this loss of free ammonia was equivalent to 2.3% of the added nitrogen.

A practical example of the effect which different sources of ammonium fertilizers can have on plant growth is shown in Figure 5.6. Tomato seedlings were grown during winter in peat–sand mixes having a pH of 6.0 (extracted in 0.01 M $CaCl_2$) before the various nitrogenous fertilizers had been added. One week later the pH values were: mix with ammonium carbonate pH 6.50, mix with urea pH 6.54, mix with ammonium sulphate pH 6.08. Seven weeks after pricking-out the mean fresh weights of the plants were: in mix with ammonium carbonate 0.49 g, mix with urea 1.18 g, mix with ammonium sulphate 5.54 g. Plants grown with calcium nitrate

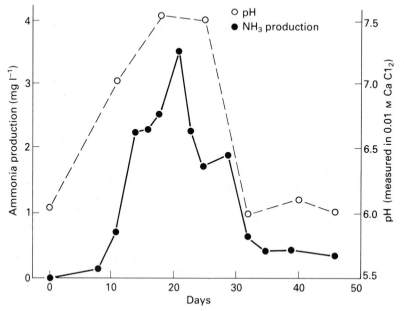

Figure 5.5 The production of free ammonia from a mix made with organic nitrogen follows the change in the pH of the mix.

Figure 5.6 Tomato seedlings grown during winter in mixes with equal amounts of ammoniacal nitrogen from three sources: *left*, ammonium carbonate; *centre*, urea; *right*, ammonium sulphate.

as the nitrogen source (not shown in the illustration) had a mean fresh weight of 8.49 g.

Nitrite As well as the direct effects that high concentrations of ammonium and free ammonia can have on plants, one of the indirect effects is to inhibit the conversion of nitrites to nitrates. Normally ammonium is quickly converted to nitrate nitrogen by the nitrifying bacteria, *Nitrosomonas* and *Nitrobacter*. When organic fertilizers in potting mixes are mineralized, localized ammonia concentration gradients can develop in the soil solution. The ammonia is oxidized to nitrite, but the nitrite is not oxidized to nitrate because *Nitrobacter* is inhibited by ammonia. The nitrite accumulation, which can be toxic to plants, will be greatest in media with low cation exchange capacity (Smith 1964). Paul & Polle (1964) showed that the growth of lettuce in an alkaline soil was inhibited at nitrite concentrations above 5 p.p.m. Birch & Eagle (1969) found that although the reduction in growth of antirrhinums could not be related to the ammonium concentration in the potting media, growth was inversely proportional to the amount of nitrite present. There was a sharp fall in the dry weight of the shoots above 20 p.p.m. nitrite nitrogen in the medium.

High levels of nitrite nitrogen frequently occur when urea has been used in the base fertilizer. Bunt & Adams (1966b) found the nitrite nitrogen level in a peat–sand mix made with urea rose to 147 mg l^{-1}; this was equivalent to 50% of the mineral nitrogen in the mix. When hoof and horn was used as the source of base fertilizer nitrogen, the maximum recorded level of nitrite nitrogen did not exceed 8% of the mineral nitrogen level. Although nitrites accumulate more rapidly under alkaline conditions, it is generally agreed that nitrite is more toxic to plants under acidic conditions. With field-grown crops, the greater part of the nitrogen absorbed by the plant will be in the nitrate form; any ammonium supplied in relatively infrequent fertilizer dressings is rapidly converted into nitrates. With pot-grown plants, however, a much greater proportion of the nitrogen is taken up as ammonium. This is mainly due to:

(a) steam pasteurization causing a temporary delay in the production of nitrites, or
(b) the naturally slower rate of nitrification in peat mixes, and
(c) the production of ammonium from slow-release forms of fertilizers, or
(d) the use of liquid fertilizers containing ammonium. Some crops are given liquid fertilizers as frequently as twice per day.

5.2.2   Mineral nitrogenous fertilizers

Although there are many mineral nitrogenous fertilizers, in practice only a few are in general use in base fertilizers for potting composts.

Ammonium sulphate $[(NH_4)_2SO_4]$ contains 21% nitrogen, all in the ammonium form. In common with most fertilizers which supply ammonium, this fertilizer is acid-forming and its use results in a fall in the media pH. Under agricultural conditions, it is generally considered necessary to apply an equal weight of calcium carbonate in order to maintain the pH level.

Ammonium nitrate (NH_4NO_3) 35% nitrogen, supplies equal amounts of ammonium and nitrate nitrogen. In its earlier form, this fertilizer was very deliquescent and had to be stored under dry conditions. To improve its handling and storage properties it is now sold in 'prilled' form. During manufacture, the concentrated ammonium nitrate solution, which has about 2% of magnesium nitrate added as an internal desiccant, is sprayed to form small droplets which are cooled and solidified by convection into prills. Ammonium nitrate is classified as an oxidizing agent, but it will not burn on its own; only if there is a fire within a confined space will ammonium nitrate increase the fire and explosive risk. *Under normal conditions of usage this fertilizer is quite safe and does not present any risk.*

Calcium nitrate   There are two types of calcium nitrate used in horticulture. Usually a small, prilled material $[5Ca(NO_3)_2 \cdot NH_4NO_3 \cdot 10H_2O]$ containing 15.5% nitrogen and 19% calcium is used in potting mixes as a base fertilizer and also for making liquid feeds. However, this material has about 8% of its nitrogen as ammonium nitrogen, which has been shown to be detrimental to early crops of tomatoes grown in NFT (nutrient film technique – a form of hydroponics). Therefore, for this purpose a hygroscopic, white, crystalline solid $[Ca(NO_3)_2 \cdot 4H_2O]$ containing 12% nitrogen and 17% calcium is used.

Urea $[CO(NH_2)_2]$ contains 46% nitrogen. From its formulation it can be considered as being an organic source of nitrogen; however, in practice it behaves as an inorganic fertilizer, being hydrolyzed in the media within a few hours to ammonium carbonate by the enzyme urease. Urea should not contain more than 2.5% of biuret, which at high concentrations is toxic to plants. Hazards associated with the use of this fertilizer are free ammonia and nitrites.

Calcium-ammonium-nitrate   To comply with legislation regarding the handling of potentially explosive materials, this fertilizer is sometimes used in place of ammonium nitrate. It is made by mixing ammonium nitrate with calcium carbonate. The analysis and the name of the fertilizer can vary with the country of origin. 'Nitrochalk' contains 15.5% nitrogen and the name, 'calcium-ammonium-nitrate', is reserved for mixtures containing 21 or 27% nitrogen, 'Cal Nitro' and 'Nitrolime' are names used for lower grade materials. If it is intended to, use these fertilizers in loamless media, it is essential that both the nitrogen and calcium contents are first checked.

Potassium nitrate (KNO_3) contains 13% nitrogen and 38% potassium. The high potassium content means that it is primarily regarded as being a potassium fertilizer.

With all these fertilizers the nitrogen is in an immediately available form; this has the advantage that the amount of mineral nitrogen present in the mix is under direct control and can be regulated according to the plant's requirements. There is also the disadvantage of increasing the salinity of the mix together with the greater risk of loss by leaching. To avoid these disadvantages and to allow more nitrogen to be added in the base fertilizer, it is common practice to use some 'slow-release' forms of nitrogen.

5.3   SLOW-RELEASE FORMS

This term is used to describe fertilizers which release nutrients slowly to plants over an extended period. These fertilizers can be classified into three groups depending upon their composition and mode of release, namely (a) organic, (b) compounds having slow rates of mineralization or dissolution, (c) coated fertilizers.

5.3.1   Organic

With traditional organic fertilizers, such as hoof and horn (13% nitrogen) and dried blood (10–13% nitrogen), mineralization of the organic nitrogen is by fungal and bacterial decomposition. The various steps in the mineralization process can be represented by the following simplified stages:

Amino acid production

organic nitrogen
in fertilizers
and $\qquad\xrightarrow{\text{by enzymes}}$ acid nitrogen + CO_2 + energy
protein nitrogen $\qquad\qquad$ (NH_2)
in organic matter $\qquad\qquad\qquad\qquad\qquad\qquad$ (I)

free amino

Ammonification

$$R-NH_2 + HOH \rightarrow NH_3 + R-OH + Energy$$
(R can be a hydrogen atom, short carbon chain or methyl group.) (II)

Stages (I) and (II) are performed by heterotrophic micro-organisms that rely on organic substances for their energy. These organisms are unable to utilize carbon dioxide and inorganic compounds.

The ammonia (NH_3) produced in stage (II) combines with the carbonic acid and other soil acids to form the ammonium (NH_4^+) ion. Only under neutral or alkaline conditions will significant amounts of ammonia persist as free ammonia.

Nitrification

$$2NH_4^+ + 3O_2 \rightarrow 2NO_2^- + 2H_2O + 4H^+ + energy \qquad (III)$$

$$2NO_2^- + O_2 \rightarrow 2NO_3^- + energy \qquad (IV)$$

Stage (III) is performed by *Nitrosomonas* and stage (IV) by *Nitrobacter*. Both of these bacteria are chemoautotrophs, i.e. they obtain their energy for growth from the oxidation of inorganic compounds, and their carbon by assimilation of carbon dioxide. The nitrifying process requires molecular oxygen and will only proceed rapidly in well aerated media. It will also be seen that hydrogen ions (H^+) are released during the nitrification process, and the continued use of organic or ammonium fertilizers will lead to the acidification of the mix.

In peat-based mixes, the micro-organisms which first break down the organic nitrogen to ammonium are more active than those which convert ammonium to nitrite and then to nitrate. This leads to a build-up of ammonium which under certain conditions can cause the pH to rise by as much as one whole pH unit. If the pH exceeds the neutral point there is the risk of free ammonia toxicity and of microelement deficiencies, primarily of boron. This situation can persist until the pH drops with the conversion of the ammonium to nitrates. The related changes in the forms of nitrogen and the pH of the mix, which occur during the mineralization of organic nitrogen, are shown in Figure 5.7. Factors which increase the rate of mineralization of hoof and horn are a reduction in its particle size, an increase in the temperature of the medium up to 40 °C and an increase in the water content until it approaches the 'container capacity' level.

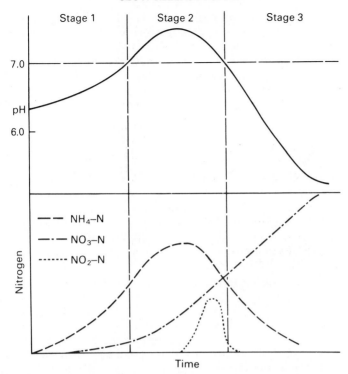

SLOW-RELEASE FORMS

Stage 1   Stage 2   Stage 3

7.0

pH

6.0

— — NH$_4$–N

— · — NO$_3$–N

· · · · · · NO$_2$–N

Nitrogen

Time

Figure 5.7 Diagrammatic representation of the changes which occur in the forms and amounts of mineral nitrogen, and in the pH when organic-type fertilizers are used in peat-based mixes. *Stage 1*: ammonium nitrogen produced, pH may rise to the neutral point. *Stage 2*: more ammonium produced, pH rises above the neutral point and free ammonia is present; nitrites may also be present; nitrates start to form. *Stage 3*: ammonium falls, nitrates increase and the pH is now below the original value.

Under glasshouse conditions of high media temperatures and available water supply, about 70% of the nitrogen can be mineralized within 30 days. Because of their relatively quick rates of release under these conditions and the high risk of toxicities, these fertilizers are not as widely used in loamless mixes as they were in the mineral soil-based John Innes composts.

5.3.2   Compounds with low rates of mineralization or dissolution

Urea-formaldehyde (UF)   Originally developed by workers in the United States Department of Agriculture, this fertilizer is formed as a controlled condensation product of urea and formaldehyde. For fertilizer purposes, the molar ratio of urea to formaldehyde is between 1.2 and 1.4 and the product has at least 35% nitrogen; most of the commercial UF fertilizers contain 38% nitrogen. Long & Winsor (1960) investigated the rates at which different urea-formaldehyde compounds were mineralized in soil, and found that methylene-diurea and dimethylene-triurea both mineralized too rapidly to be of value as slow-release nitrogen sources, whereas tetramethylene-pentaurea has too slow a rate of decomposition. Trimethylene-tetraurea had the most useful mineralization rate of all the compounds tested. Commercial UF fertilizers consist of mixtures of these compounds with some free urea also present and it is customery to state their 'availability index' (AI) as a means of assessing their probable rates of mineralization in soils:

$$\text{AI} = \frac{\begin{array}{c}\text{\% cold water insoluble nitrogen}\\ -\text{ \% hot water insoluble nitrogen}\end{array}}{\text{\% cold water insoluble nitrogen}} \times 100$$

This index approximates to the amount of insoluble nitrogen which nitrifies in an average soil in about six months; for use in potting mixes the fertilizer should have an AI value of between 40 and 45.

Urea-formaldehyde is mineralized in media by bacterial action, with some chemical hydrolysis also occurring in the initial stages. The rate of mineralization is not significantly affected by the particle size of the fertilizer but is increased by high temperatures and by low pH values (Winsor & Long 1956). This effect of pH on the mineralization rate is in contrast to that found with hoof and horn; comparison of the mineralization rates of these two fertilizers in peat–sand mixes at two pH values is shown in Figure 5.8. It has been estimated that approximately one third of the nitrogen present in urea-formaldehyde is available to plants within a few weeks, a further third within several months and the remaining third within 1–2 years. For this reason, it is unlikely that the average pot-grown plant recovers more than 50% of the applied nitrogen. With border-grown crops, however, some of the remaining nitrogen would be recovered by successive crops. Commercial urea-formaldehyde fertilizers available in Britain are 'Nitroform' and 'Ureaform'.

Crotonylidene-diurea (CDU) German workers have developed a fertilizer based upon the condensation of urea with crotonaldehyde. This crotonylidene-diurea has the trade name 'Crotodur' and contains 28% nitrogen, 90% of which is in a slow-release form and the remaining 10% is present as nitrate nitrogen. 'Crotodur' forms the basis of such slow-release fertilizers as 'Triabon' (16–8–12–4; N, P_2O_5, K_2O, MgO). Nitrogen release is by bacterial and fungal breakdown and, as with urea-formaldehyde, the decomposition rate is inversely related to the pH. Very little nitrogen is released during the first six weeks if the pH is about 6.0. The rate of mineralization is also dependent upon the soil temperature: at the low soil temperatures of 7–10 °C virtually no nitrogen is released. A reduction in the particle size of the fertilizer also gives a higher rate of mineralization. Results of incubation tests in peat–sand mixes at the Glasshouse Crops Research Institute have shown this material to have the slowest release rate of all the slow-release fertilizers tested, it is widely used in peat media in Germany.

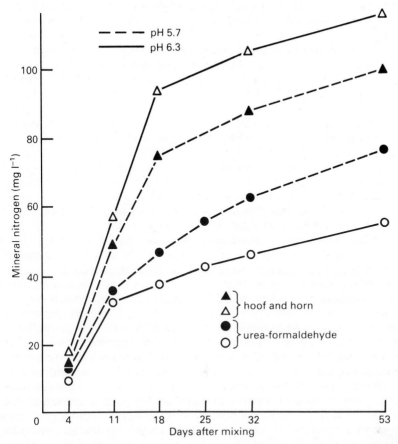

Figure 5.8 Contrasting the effect of the mix pH on the mineralization rates of hoof and horn and urea-formaldehyde. With hoof and horn the high pH gave the fastest rate of mineralization, whereas with urea-formaldehyde the fastest rate of mineralization occurred at the low pH.

Isobutylidene-diurea (IBDU) This material is a condensation product of urea and isobutyraldehyde in a 2:1 mole ratio; it has 32% nitrogen and a very low solubility in water. Nitrogen release is by dissolution and hydrolysis and is not significantly affected by microbial activity. The principal factor known to increase the rate of hydrolysis is a decrease in the particle size and hardness: the normal size of material used in mixes is 0.7–2.0 mm. Hydrolysis is also more rapid at low pH; mixes containing IBDU which were kept at pH 5.7 for four weeks had twice as much nitrogen as similar mixes maintained at pH 7.7. An increase in soil temperature has only a relatively small effect on the release rate. It can normally be safely used at high rates of application but a mild toxicity and some depression in growth has been observed when used at the rate of 500 mg nitrogen l^{-1} of peat, equivalent to 1.56 g of fertilizer l^{-1}.

IBDU is distinct from UF and CDU in that the release mechanism is chemical rather than microbiological.

Magnesium-ammonium-phosphates This group of fertilizers were originally regarded as a slow-release source of nitrogen. A low pH and high moisture content increase the rate of release; temperature has only a very slight effect on the rate of dissolution. The principal method of controlling the release is by regulating the particle size; small particles have a much higher rate of release because of their higher surface : volume ratio. Release rates have been found to be greater than would be expected purely from the water solubility values, and it is concluded that diffusion of the ammonium away from the particle surface and also the conversion of the ammonium to nitrate are responsible for the higher rate of dissolution. The grade of material suitable for use in mixes has 80% of the particles in the 1.5–3.0 mm range; larger particles give less uniformity of fertilizer distribution. This material is available in the USA under the trade name 'MagAmp', it has an analysis of 7–40–6 + 12; N, P_2O_5, K_2O, Mg. In Britain a product known as 'Enmag' contains 6% N, 20% P_2O_5, 10% K_2O and 8.5% Mg; the potassium is present as potassium sulphate.

Two effects sometimes associated with magnesium-ammonium-phosphates have been potassium deficiency, induced by the high rates of magnesium, and an iron deficiency, induced by the high rates of phosphorus forming insoluble iron phosphates. Potassium deficiency is controlled by increasing the amount of potassium in the base fertilizer, and the iron deficiency can be controlled by applying a 0.1% solution of iron chelate as required. The value of these materials as slow-release sources of nitrogen has generally been overemphasized; their low nitrogen and high phosphorus content does not commend them for this purpose. The high rate at which the water-soluble phosphatic fertilizers are lost from peat–sand mixes by leaching rather suggests that the greatest potential use for the metal-ammonium-phosphates is as slow-release sources of phosphorus.

Oxamide Oxamide is a diamide of oxalic acid with 31.8% nitrogen. Its dissolution rate is controlled by the particle size; large particles give a slower initial release rate, but this is sustained for a longer time. The recovery of nitrogen by plants has been highest in soils with high pH values. Although this material has given good results in experiments, it is not generally available owing to its high cost of production.

5.3.3 Coated fertilizers

Sulphur-coated urea This type of fertilizer has been developed in the USA by the Tennessee Valley Authority and by Imperial Chemical Industries in Britain. The American product, known as SCU, has 38.1% nitrogen with the coating forming 20% of the total weight, 16% being sulphur, 3% wax sealant and 1% a conditioner; there is also a microbicide incorporated, to protect the sealant against rapid microbial decomposition. Sulphur-coated urea can be prepared with different dissolution rates. Furuta *et al.*

(1967) used materials having dissolution rates of 6, 5 or 1% per day, and found the two materials with the highest rates of dissolution to be toxic to some ornamentals. They recommended the material with the 1% dissolution rate be used at 100 g nitrogen m^{-3} of media (0.26 kg m^{-3} of fertilizer). The dissolution rate measures the rate at which the nitrogen is released under standard laboratory conditions and is defined as:

$$\text{dissolution rate} = \frac{\text{wt of fertilizer dissolved in water at 38\,°C in 24\,h} \times 100}{\text{original weight of fertilizer}}$$

The British product is known as 'Gold-N' and contains 32% nitrogen. As with the American product, it consists of urea which is coated with sulphur and a sealant, but it does not contain a microbicide. No direct comparisons of the dissolution rates of the two products appears to have been made, but Prasad & Gallagher (1972) concluded that the American material mineralized too slowly in peat to be a suitable source of nitrogen for tomato nutrition; in eight weeks the sulphur-coated urea had released only slightly more mineral nitrogen than had urea-formaldehyde. Working with the British product 'Gold-N', Bunt (1974a) found this had a significantly faster release rate than urea-formaldehyde, comparable values of mineral nitrogen after four weeks being 200 and 126 mg nitrogen l^{-1} respectively. This suggests that the American product has a slower dissolution rate than 'Gold-N'.

Coated NPK fertilizers There are also a number of resin- or polymer-coated fertilizers that supply phosphorus and potassium as well as nitrogen, e.g. 'Osmocote', 'Nutricote', 'Ficote'. The characteristics and use of these fertilizers are discussed in section 6.6.1.

5.4   CHOICE OF FERTILIZER TYPE

Several factors must be considered when deciding whether to use inorganic fertilizers or the slow-release types in loamless mixes. The most important points for consideration will be:

(a) the probable length of the period between mixing and using, i.e. the storage period,
(b) the type of crop and season,
(c) the ability to give liquid fertilizers and the relative convenience of the two systems,
(d) the relative costs of the various fertilizers.

5.4.1   Storage of mixes

The risk of phytotoxicity occurring when plants are grown in mixes that have been stored depends upon a number of factors, the main ones being: (a) the form and rate of nitrogen used, (b) storage conditions that can affect release or mineralization rates and (c) the environment in which plants are grown using the stored mix.

Mixes made with inorganic nitrogen usually show only slight changes during storage; mainly a slow conversion of ammonium to nitrate nitrogen and, with some peats, a reduction in the mineral nitrogen (see Figs 5.12 and 5.13) with an accompanying increase in organic nitrogen. High media temperatures and moisture contents increase the rates of biological conversion and fixation of nitrogen. If mixes with coated fertilizers are stored moist and in a warm situation, sufficient nutrients can be released to cause typical high salinity damage to plants. Mixes with organic-type nitrogenous fertilizers are the most prone to storage problems. A feature of mixes made with young sphagnum peats has been that although fertilizers such as hoof and horn and urea-formaldehyde are mineralized to ammonium quite rapidly, the subsequent conversion of ammonium to nitrate nitrogen is slow, thereby resulting in possible toxicities from ammonium, high pH, free ammonia and microelement unavailability.

This situation is clearly seen from the work of Bunt & Adams (unpublished), where the interaction of the storage period and the source of nitrogen was investigated by preparing two mixes, one with hoof and horn, and the other with calcium nitrate. After preparation, the two mixes were stored, and samples used over nine successive weeks to grow tomato seedlings whose growth rates were recorded from weekly measurements of the leaf lengths. The effect which the length of the storage period had on plant growth has been summarized in Figure 5.9. The total length of the leaves of plants in the hoof and horn mix after an eight-week growth period has been expressed as a percentage of the total leaf length of comparable plants grown in the calcium nitrate mix. Also included in Figure 5.9 are the ammonium, nitrate and pH levels of the stored hoof and horn mix at the time it was used to prick out the tomato seedlings. There was virtually no change in the stored mix made with the calcium nitrate.

When used without any storage period, plants in the hoof and horn mix were only very slightly inferior to those grown in the calcium nitrate treatment. As the storage period was increased, there was a progressive decrease in the growth of the plants in the hoof and horn mix. After three weeks of storage, plants growing in the hoof and horn mix made only 10% of the growth of plants in the corresponding calcium nitrate mix. As the storage period was further increased, the growth of plants in the hoof and horn mixes steadily improved in relation to the corresponding calcium nitrate mixes, until at six weeks after mixing growth was essentially similar in both mixes. It will be seen that, at this time, the ammonium and pH levels in the stored mixes were falling and nitrates were increasing. For experimental reasons the mixes were stored moist and warm; at lower temperatures and moisture levels mineralization would have been slower. These results illustrate a principle rather than define when stored mixes are 'safe' or 'unsafe' to use.

The availability of boron in the mix would also have increased as the pH fell. Nitrification occurs more rapidly in mixes made with sedge peat than in sphagnum peat mixes (see Fig. 5.12). The problem is also less severe in

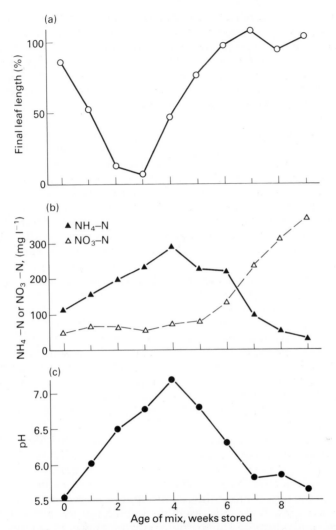

Figure 5.9 Effect of the form of nitrogen and duration of mix storage on the growth of tomato seedlings in winter. (a) Growth of plants in mixes made with hoof and horn, and stored for varying periods, expressed as a percentage of the growth of plants in mixes made with nitrate nitrogen and stored for similar periods. Growth was assessed by the total length of the leaves eight weeks after pricking-out the seedlings. (b) The change in the ammonium and nitrate nitrogen levels of the hoof and horn mixes during storage. (c) The change in the pH of the hoof and horn mixes during storage.

sphagnum peat that has already grown a crop as the greater population of *nitrobacter* present in the peat gives more rapid nitrification.

Where there is a risk that mixes containing urea-formaldehyde will be stored before use, the maximum rate of urea-formaldehyde should not exceed $0.5\,\mathrm{kg\,m^{-3}}$. If a medium containing too much slow-release nitrogen has inadvertently been used, the nitrogen content cannot be reduced easily. Although both the ammonium and nitrate forms of nitrogen can be readily leached, other nutrients will also be removed, and excessive watering of plants in non-porous containers in winter is better avoided. One approach to the problem of high levels of nitrogen in media, caused by the continued release of nitrogen from slow-release fertilizers in excess of the plants' requirements, has been to apply a solution of sucrose. This changes the carbon:nitrogen ratio, and some of the mineral nitrogen is temporarily immobilized. Solutions of 1 kg of sucrose per 30 l of water (2 lb per 6 gal) have been applied to geraniums at the rate of 150 cm^3 per 10-cm pot (6 fl oz per 4-in pot) with beneficial results.

Phytotoxicity symptoms   A range of symptoms have been observed in mixes having organic-type nitrogen, these include: root damage, iron chlorosis, epinastic curling of leaves, restricted growth or death of apices – often followed by multiple branching, leaves fleshy and brittle, cupping of leaves and distorted growth of the lamina. Species vary in their susceptibility to the problem and in the intensity and range of symptoms seen, they also differ in their ability to resume normal growth.

Some of the symptoms are directly associated with high ammonium levels, e.g. epinastic curling in tomato leaves. Often symptoms are indirectly related to high ammonium, e.g. a rise in pH of the media (see Fig. 5.9) resulting in an induced boron deficiency (see Fig. 7.2); with nitrification the pH falls and boron availability is increased. Although the availability of other microelements is also affected by the pH, boron is the only element that has been positively identified with ammonium-induced deficiencies. High ammonium levels also reduce calcium uptake, and affected plants often have lower calcium in the leaf tissue (Tew Schrock & Goldsberry 1982).

Certain symptoms have not been positively related to either deficiencies or excess of minerals in the tissue, they have resembled virus or hormone damage symptoms. Recently English & Barker (1983) have described such symptoms in tomato plants grown in a sludge compost treated with the nitrification inhibitor nitrapyrin. They concluded that the distorted leaf growth did not resemble ammonium toxicity or nutrient deficiency, and attributed it to the action of nitrapyrin or one of its metabolites as a growth regulator. Examples of the distorted growth observed with brassicas, tomato and lettuce grown with organic-type nitrogenous fertilizers are shown in Figure 5.10. Seedlings grown in winter and early spring appear to be more susceptible to the problem than those grown later in the season.

Microelement availability   When mixes are made with organic forms of nitrogen and then stored for a period before being used, the availability of the microelements is reduced because the build-up of ammonium causes a rise in the pH. The significance of this for the growth of plants in stored mixes is shown in Figure 5.11. By comparison with plants grown in the fresh mixes, those grown in the stored mixes had made very little growth.

Figure 5.10 Examples of injury to (a) brassicas and (b) lettuce, caused by growing in stored mixes made with organic-type slow-release nitrogen. Symptoms are usually most evident on young plants grown in winter or early spring and vary with the plant species.

Continued.

(c)

Figure 5.10 *continued* Some plants produce leaves that are thick, curled or wrinkled, others, e.g. tomato (c), may show typical symptoms of virus or herbicide damage. Affected leaves will often have lower levels of calcium and boron than normal leaves.

fresh mixes, those grown in the stored mixes had made very little growth. However, plants growing in mixes to which microelements in the form of inorganic salts had been added before storage, made normal growth. Subsequent work has shown that the temporary unavailability of boron is one of the principal factors involved.

Some of the other factors to consider in making the choice between slow-release and mineral forms of fertilizer are less easy to quantify. For example, the ability to apply fertilizers in liquid form varies widely between nurseries. Many growers of greenhouse crops rely on using mineral fertilizers in the mix, supplemented with liquid feeding, whereas most growers of hardy ornamental nursery stock use controlled-release fertilizers.

Figure 5.11 Changes in the pH and forms of mineral nitrogen in stored mixes, made with organic nitrogen, can also induce microelement deficiencies, especially of boron. *Left*: antirrhinum seedling pricked out into a freshly made mix, no microelements added. *Centre*: mix stored for three weeks, no microelements added. *Right*: microelements added during mixing then stored for three weeks before seedling pricked out.

5.5 NITROGEN AND PEAT

Peats differ from many other materials used to make loamless mixes because they contain relatively large amounts of nitrogen, usually between 1 and 2.5% of their dry weight, whereas perlite, plastics, sand and vermiculite, etc., contain little or no nitrogen. To put this apparently high nitrogen content of peat into perspective, however, the following three factors must be considered. First, the nitrogen is largely in an organic form and is therefore not immediately available. Second, although peats may appear to contain ten or more times as much total nitrogen as mineral soils, their bulk density is only about one tenth or less of that of a mineral soil. When the amount of nitrogen is considered in relation to the *volume* rather than the *weight* of peat in which the plant is growing, there is a corresponding reduction. Third, peats contain some organic matter that is fairly readily decomposed and, during the decomposition process, some of the mineral nitrogen will be used by the micro-organisms, and will not, therefore, be available to the plants. This situation is known as 'nitrogen immobilization'.

Often the carbon:nitrogen ratio is used as a means of categorizing soils and other materials with regard to their probable behaviour in the mineralization of organic nitrogen or in the immobilization of mineral nitrogen during cultivation.

5.5.1 *Carbon:nitrogen ratio (C:N)*

Quite wide differences exist between types of peat with respect to their C:N ratios: these vary from about 40:1 for some young sphagnum peats to 20:1 for sedge peats. By contrast, the mineral soils of Britain have values nearer to 10:1. However, too much reliance must not be placed upon the C:N ratio as giving a simple measure of the availability of nitrogen in various materials; much will depend upon the form of the organic matter. For example, Bollen (1953) found a very high C:N ratio of 729:1 for Western Red Cedar sawdust, but, because the material contained 60–70% of lignocellulose which decomposes very slowly, less nitrogen was required to offset the lock-up of nitrogen by the microorganisms than would be indicated from the C:N ratio. In practice, it is found that the amount of carbon left after a period of decomposition is primarily determined by the chemical composition of the material, rather than by the C:N ratio; materials high in lignins decompose very slowly, while hemicellulose and cellulose decompose rapidly. Puustjärvi (1970a) found that the amount of easily decomposable organic matter in various peats ranged from 65.6% for *Sphagnum fuscum* to 15.2% for a *Bryales–Carex* peat. The nitrogen requirements of *S. fuscum* peats will therefore be greater than that of sedge peats and, during the early stages of decomposition, an increased rate of nitrogen supply will be required to prevent the plants showing nitrogen deficiency symptoms. Later, when the rate of decomposition is reduced, some of this immobilized nitrogen will be released to the plant upon the death of the micro-organisms. This immobilization of nitrogen and its subsequent release will have the greatest effect with border crops which are grown in relatively large volumes of media, and where liquid feeding is not of such importance as it is with pot-grown crops.

Some results obtained at the Glasshouse Crops Research Institute, on nitrogen studies with different types of peat, are shown in Figures 5.12 and 5.13. Known amounts of ammonium sulphate together with other base fertilizers supplying phosphorus, potassium, calcium, and magnesium and microelements, were added to three peats, i.e. sphagnum A, a young peat, H2 on the von Post scale, sphagnum B, a more humified peat, H4, and also a sedge peat. These were moistened, placed in an incubator and ammonium and nitrate levels were determined at intervals.

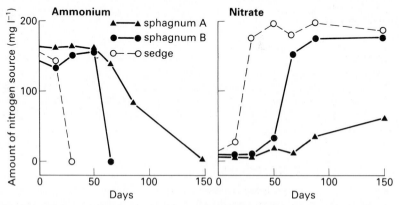

Figure 5.12 Comparative rates of conversion of ammonium sulphate to nitrate nitrogen in three peats. Sedge peat showed the most rapid rate of nitrification, and sphagnum peat A the slowest rate of nitrification.

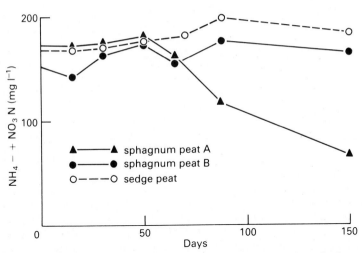

Figure 5.13 Immobilization of nitrogen in peats. In both the sedge peat and sphagnum peat B, there was no reduction in the amount of water soluble nitrogen, but in sphagnum peat A approximately 50% of the nitrogen was immobilized after 150 days.

As seen in Figure 5.12, conversion of the ammonium into nitrate nitrogen was fairly rapid in the sedge peat, and slower in the sphagnum peats; sphagnum peat A being significantly slower than sphagnum peat B. Although none of the peats had been steam sterilized, the slow formation of nitrates in the sphagnum peats is similar to the delay found in mineral soils after steam sterilization. Under glasshouse conditions, the time taken for the nitrifying bacteria to build up in population will be primarily dependent upon four factors: (a) the temperature of the medium, (b) its moisture content, (c) its pH and (d) the degree of bacterial inoculation. The most important source of inoculation is the carry-over of inoculum on the roots during pricking-out or repotting from a medium in which the nitrifying bacteria have already built up a large population.

In Figure 5.13, the total amount of mineral nitrogen, i.e. the ammonium plus nitrate forms, present in the three peats during the course of the experiment shows that, after day 50, a significant amount of the mineral nitrogen in sphagnum peat A had been immobilized. No reduction in the mineral nitrogen levels had occurred in either sphagnum peat B or the sedge peat. Results from this and other experiments show that usually the ammonium form of nitrogen is preferentially utilized by the bacteria in decomposing the peat.

CHAPTER SIX

Other macroelements

In addition to nitrogen, there are a number of other elements which are essential to plant growth; those required in relatively large amounts are known as *macro-* or *major* elements, and those required only in relatively small amounts are known as *micro-* or *trace* elements. Irrespective of the actual quantities required by the plants, however, both groups are equally *essential*; only a few elements such as barium, fluorine, iodine, and strontium do not appear to be essential for plant growth. The essential macroelements are nitrogen, phosphorus, potassium, calcium, magnesium and sulphur.

Most of the materials described in Chapter 2 as being suitable for making loamless mixes are deficient in several or all of the macroelements, by comparison with a fertile mineral soil; two exceptions being peat, which has a relatively high organic nitrogen content, and vermiculite, which contains large amounts of potassium and magnesium. Relatively large quantities of fertilizers may therefore be required in order to raise the nutrients to suitable levels, if these materials are used to make potting mixes.

The actual amount of fertilizers required to prepare potting mixes will depend upon three main factors:

(a) The amount of nutrients already present in the bulk materials and the rate at which they become available to the plant.
(b) The behaviour of the fertilizers when added to the mix; do they remain largely in a water-soluble form or are they liable to fixation and then become unavailable? Does the material have a significant cation exchange capacity?
(c) The species of plant to be grown; has it a short growing period and a high nutrient demand, e.g. pot *Chrysanthemum*, or does it have a low demand for nutrients, e.g. *Calceolaria*? Is there a known requirement for a specific element, e.g. aluminium to produce blue flowers in *Hydrangea*, or is there likely to be a specific deficiency or toxicity, e.g. molybdenum deficiency in lettuce and boron toxicity in *Beloperone*?

Not only is it necessary to maintain a supply of each of the plant nutrients, but a balance must also be maintained between the concentrations of the different elements. The effects of individual elements are often discussed separately but large and important *interactions* occur between nutrients. High levels of some nutrients may accentuate deficiencies of other nutrients. For example, the application of nitrogen and potassium may cause an increased rate of growth, resulting in phosphorus deficiency, or the addition of large amounts of potassium to the mix may reduce the plants' uptake of magnesium, and although a reasonable quantity of magnesium is present in the mix, magnesium deficiency can still occur. This is known as a potassium-induced magnesium deficiency and is one example of antagonism between elements. These and other nutrient interactions are readily seen from factorial design experiments which allow both the main effects and the interactions to be observed. Experiments of this type yield much more information than simple trials, and their use has been advocated by R. A. Fisher (1926) who said, 'No aphorism is more frequently repeated in connection with field trials than that we must ask nature a few questions, or, ideally, one question at a time. The writer is convinced that this view is wholly mistaken. Nature, he suggests, will best respond to a logically and carefully thought out questionnaire, indeed if we ask her a single question, she will often refuse to answer until some other topic has been discussed.'

The functions in plant nutrition of the remaining five macroelements and the form in which they can most conveniently be applied will now be examined, together with some of the most important of the interactions between the elements and the environment. The modern practice of referring to nutrients by element rather than oxide, e.g. K instead of K_2O, has been adopted in this book. The traditional oxide basis is often illogical, e.g. when stating the K_2O content of KCl, the oxide basis has only been retained when referring to other published work. Factors for converting from the element to the oxide basis are included in Appendix 7.

6.1 PHOSPHORUS

Chemical analysis of the tissue of a typical pot-grown plant will show that, by comparison with nitrogen and potassium, the amount of phosphorus absorbed is relatively low, often only one tenth of the amount of either of the other elements. This does not mean that less attention is required in the phosphorus nutrition of pot plants; indeed, maintaining an adequate and continuing supply of phosphorus in loamless mixes is an important aspect of macroelement nutrition.

Phosphorus is present in relatively large quantities in the protoplasm and nucleus of the cell. It is a component of the nucleic acids and sugar phosphates and is essential for many of the energy transfer processes, such as photosynthesis and the breakdown of carbohydrates, which occur in plants. Deficiency symptoms can include a dark green colour of the upper leaves, and the lower leaves turning a pale green with yellow patches developing at a later stage. Phosphorus is one of the mobile elements within the plant, and, when there is a deficiency in the medium, the plant is able to withdraw some of the phosphorus from the lower leaves and transfer it to the young, actively growing tissue. This is the reason why deficiency symptoms are first seen in the older leaves. Young tomato seedlings, deficient in phosphorus, will often show a reddish-purple colour on the under-sides of the cotyledons. This pigmentation can be caused by low air temperatures, but it is also known that low soil temperatures have a greater effect on reducing the absorption of phosphorus by the plant than on the absorption of potassium and other cations. High levels of available phosphorus in media are traditionally considered necessary for good root growth and the rapid establishment of young seedlings.

In mineral soils, the concentration of phosphorus in the soil solution is low by comparison with other macroelements. It has been estimated that in order to maintain adequate growth rates the phosphorus in the soil solution has to be renewed several times each day to replace that taken up by the plants. Unlike mineral soils, which contain apatite and other minerals that provide a natural supply of phosphorus, the materials commonly used to make loamless mixes do not have a natural supply of phosphorus, and neither do they have the ability to fix or retain phosphorus to the same extent as mineral soils. Materials used to make loamless mixes contain relatively little aluminium and iron, and their capacity to fix phosphorus in the insoluble aluminium and iron forms can, from the pot-plant grower's point of view, be regarded as insignificant. More attention must therefore be given to maintaining adequate levels of phosphorus in these mixes, than is the case with mineral soil composts. The high degree of water solubility of phosphorus in peat–sand mixes, in contrast to that of a John Innes compost, was discussed in Chapter 4.

The chemical form in which phosphorus is present in media is largely determined by the pH. Olsen (1953) concluded that, at pH 5, practically all the inorganic orthophosphorus present in the soil solution was in the form of $H_2PO_4^-$, and, as the pH was raised to 7.2, the phosphorus was equally divided between the $H_2PO_4^-$ and HPO_4^{2-} forms. High pH values reduce the availability of phosphorus, and plants grown at a high pH contain much less phosphorus than those grown at a lower pH. The total amount of mineral phosphorus present in peats is low, being about 0.01–0.05% of the oven-dry weight; the amount of organic phosphorus which is mineralized and made available during the growth period of a typical pot-plant is also small. For all practical purposes, it must therefore be assumed that all of the plant's requirement for phosphorus must be met by the application of phosphatic fertilizers.

Although several phosphatic fertilizers are available, in practice only a few of them are used in preparing loamless mixes and it is convenient to consider them under two groups: (a) those in which the phosphorus is water-soluble and (b) fertilizers having insoluble forms of phosphorus. It should be noted, however, that not all of the so-called water-soluble fertilizers are suitable for making liquid fertilizers. Although the phosphorus may be in a water-soluble form, other non-soluble compounds may be present, e.g. the calcium sulphate in superphosphate.

6.1.1 Water-soluble phosphatic fertilizers

Superphosphate This is the most widely known and used phosphatic fertilizer. It is produced by adding sulphuric acid to finely ground rock phosphate. This gives a water-soluble, monobasic calcium phosphate mixed with twice its weight of calcium sulphate. It normally has 8.72% phosphorus (20% P_2O_5), 85% being water-soluble, the remainder being insoluble rock phosphate; it also has 18% calcium and 12% sulphur with some microelement impurities such as boron and zinc. The fertilizer is available in powder or granular forms.

When the nitrogen in the base fertilizer is supplied in organic form, high rates of superphosphate often show marked beneficial effects on plant growth which cannot be accounted for simply on the grounds of increased levels of available phosphorus and an improved $N:P$ ratio; this situation is demonstrated in Figure 6.1. When nitrogen was supplied in the nitrate form, plant growth responded little to varying the amount of superphosphate; when the nitrogen was supplied as urea, however, there was a progressive increase in growth as the amount of superphosphate was increased. Although superphosphate is regarded as having a neutral effect in agricultural soils, when used in a peat–sand mix it will reduce the pH (Table 6.1). Bunt & Adams (1966b) investigated the beneficial effect of using high rates of superphosphate, when the nitrogen source was hoof and horn, by following the changes in the forms of nitrogen and the mix pH over a period of eight weeks. They concluded that the probable reasons for the beneficial effect of the superphosphate on plant growth were the reduction in the mix pH, which also slowed down the rate of organic nitrogen mineralization, a 50% reduction in the amount of free ammonia in the mix, and also a reduction in the amount of nitrite nitrogen produced.

Table 6.1 Effect of forms and rates of superphosphate on the mix* pH, determinations made in 0.01 M $CaCl_2$. (Figures in parentheses are for determinations made in water.)

| Fertilizer | Rate (kg m^{-3}) | | | |
	nil	0.75	1.50	3.00†
single superphosphate	5.22	4.87	4.57	4.39
(8% phosphorus)	(5.71)	(5.28)	(4.79)	(4.64)
triple superphosphate	5.22	4.80	4.63	4.46
(21% phosphorus)	(5.71)	(5.35)	(5.20)	(4.87)
at equivalent rates of phosphorus				

* The mix was made from 75% Irish peat and 25% fine sand. Other fertilizers were used at the normal rate for the GCRI mix (§ 8.8).

† These rates are approximately equivalent to 1 lb 5 oz, 2 lb 10 oz, and 5 lb 4 oz per cubic yard.

Figure 6.1 The interaction of rates of superphosphate and the nitrogen source. *Top* (*left to right*): decreasing rates of superphosphate with urea as the nitrogen source. *Bottom* (*left to right*): decreasing rates of superphosphate with calcium nitrate as the nitrogen source.

Triple superphosphate As the name suggests, this fertilizer contains about three times as much phosphorus as single superphosphate. It is made by using phosphoric acid in place of sulphuric acid to dissolve the rock phosphate. The fertilizer has 18–20% water-soluble phosphorus (42–47% P_2O_5) and, unlike single superphosphate, it does not contain any significant amount of calcium sulphate. It is available in both powder and granular forms.

Monoammonium phosphate ($NH_4H_2PO_4$) This fertilizer has 12% nitrogen and 26.6% phosphorus, it is known as MAP in the UK and in the USA the agricultural grade is known as 'Ammophos A'. It is made by combining ammonia with a low grade of phosphoric acid obtained from the wet process in which rock phosphate is treated with sulphuric acid. In horticulture, monoammonium phosphate is used chiefly as a phosphorus source in liquid feeds, and, to minimize the formation of precipitates, it is essential that only the high grade fertilizer (§ 9.2.6) made from thermal process phosphoric acid is used. In the thermal process of phosphoric acid manufacture, elemental phosphorus is first produced from rock phosphate in an electric-arc furnace and then oxidized to P_2O_5 with air. This method of production gives a material with very low amounts of such impurities as iron and aluminium which can cause precipitation problems. Monoammonium phosphate made from this grade of phosphoric acid has an analysis of 25% phosphorus and 12% nitrogen.

Diammonium phosphate ($(NH_4)_2HPO_4$)   This is also known as 'Diammonphos': it contains 23% phosphorus and 21% nitrogen.

Both of the ammonium phosphates are being increasingly used in high-analysis base fertilizers for agricultural purposes where their higher nutrient content and lower handling costs make them competitive with the lower analysis phosphorus fertilizers. The analysis of these two fertilizers varies slightly with the method used in their manufacture.

Phosphoric acid (H_3PO_4)   A caustic liquid containing 24% phosphorus. It is sometimes used by fertilizer manufacturers as a phosphorus source in liquid feeds and is also applied in the irrigation water for some field crops in western parts of the USA. Some care in its handling is required, if it is used by growers in the preparation of liquid feeds on the nursery. (See monoammonium phosphate, regarding suitable grades of material.)

Superphosphoric acid   This is the most concentrated form of phosphorus available, containing 33% phosphorus. It is primarily used in the formulation of commercial liquid fertilizers; approximately half the phosphorus is in the normal orthophosphoric (H_3PO_4) form, and the remainder is largely present as pyrophosphoric acid ($H_4P_2O_4$). This latter form can sequester or chelate metals such as iron, aluminium, copper, zinc and manganese which would otherwise be precipitated as phosphates and so reduce the solubility of the fertilizer. This chelating action makes it a useful source of phosphorus for liquid fertilizers.

Polyphosphate   When superphosphoric acid is neutralized with ammonia, it gives a highly concentrated water-soluble fertilizer containing 15% nitrogen and 27% phosphorus; 50% of the phosphorus is in the orthophosphate form and about 50% in the polyphosphate form, mostly as triammonium pyrophosphate [$(NH_4)_3H_4P_2O_7$]. The polyphosphates are hydrolyzed in the soil to orthophosphate, before being absorbed by plants.

6.1.2   Slow-release or insoluble forms of phosphorus

Phosphatic fertilizers are also available in insoluble forms; rock phosphate and basic slag are traditional forms of insoluble phosphate fertilizers and are widely used in agriculture. Other materials such as the metal ammonium phosphates and the metaphosphates have recently been used in loamless mixes as a means of supplying phosphorus over a prolonged period.

Basic slag   A by-product from the manufacture of steel from iron ores rich in phosphates, the slag is finely ground and contains 3–9% water-insoluble phosphorus which is largely present as tetracalcium phosphate $(Ca_4P_2O_9)$, with some silicophosphate. This wide variability in phosphorus content, together with its strong basicity, equivalent to about 60% by weight of calcium carbonate, does not generally make it suitable phosphorus supplying fertilizer for horticultural purposes; its only appeal lies in supplying phosphorus in an insoluble form.

Rock phosphates   Finely ground rock phosphates, containing 11–13% phosphorus, are another form of slowly available phosphorus. Traditionally used on grasslands, their effectiveness and composition vary with the source of the material. A fine particle size, a high organic content of the soil and a low pH are factors generally believed to be important for the utilization of this form of phosphorus.

Other traditional slow-release phosphatic fertilizers, such as bonemeal and steamed bone flour, have variable amounts of phosphorus and nitrogen and are relatively expensive.

Magnesium-ammonium-phosphates ($MgNH_4PO_4 \cdot H_2O$)   Fertilizers in this group are of variable composition. A typical analysis would be 8% nitrogen, 20% phosphorus, 13% magnesium; potassium can also be included to give a combined magnesium-ammonium- and potassium-ammonium-phosphate. Some aspects of this group of fertilizers have already been discussed in Chapter 5. The phosphorus rate of availability from this group of fertilizers is such that, for those pot plants grown over a three month period, e.g. *Chrysanthemum* and *Poinsettia*, sufficient fertilizer can be added to meet the total phosphorus requirements of the crop without the need to include phosphorus in the liquid feeds (§ 6.1.3 & Fig. 6.2).

Calcium metaphosphate ($Ca(PO_3)_2$)   Calcium metaphosphate is a glass-like material produced by burning elemental phosphorus in a combustion chamber into which finely ground rock phosphate is injected; it is tapped from the furnace as molten glass and then cooled and crushed. This gives a material having 27–28% phosphorus, about three-quarters of which is derived from the elemental phosphorus and the remainder from the rock phosphate; it also has 25% of calcium oxide. The material is only slightly soluble in water but it is sufficiently soluble in the weak acids present in the soil to give a higher rate of availability to plants.

Potassium metaphosphate (KPO_3)   Although this material contains slightly less phosphorus (25%) than potassium (32%), its prime interest lies in increasing the range of slow-release phosphorus sources. It has a low water-solubility but hydrolyzes to give the more soluble orthophosphates, the rate of hydrolysis being affected by the moisture content, the temperature and acidity of the mix.

Dicalcium phosphate dihydrate ($CaHPO_4 \cdot 2H_2O$) This contains 18% phosphorus and can be produced with a range of granule sizes. The granules are quite hard, and unlike those of triple superphosphate do not slake readily in water. It has given promising results as a slow-release phosphorus source in peat mixes.

6.1.3 Use of slow-release phosphorus fertilizers

Orthophosphates can easily be lost from peat-based mixes by leaching (§ 4.3.2); to avoid the necessity of then having to include phosphorus in the liquid feed, several fertilizers were tested for their ability to provide a continuous supply of phosphorus to plants over a long period (Bunt 1980). The standard GCRI formula was used but the superphosphate was omitted and replaced by 50 mg l^{-1} of phosphorus from various slow-release sources. Availability of the phosphorus to plants was measured by periodically analyzing the tissue of rye grass, other nutrients having been supplied by frequent liquid feeding. The potential loss of phosphorus from the fertilizers was measured by leaching fallow containers over a six month period.

The initial and total losses of phosphorus by leaching are shown in Table 6.2. Superphosphate, which was used as the reference source of phosphorus, had the highest loss of any fertilizer, most of which occurred in the first leaching. With potassium metaphosphate and magnesium ammonium phosphate the initial losses were very low, but increased over the next two months as the fertilizers slowly dissolved. Losses from the bonemeal and Gafsa rock phosphate were low throughout the experiment.

Availability of the different forms of phosphorus to the grass is shown

Table 6.2 The loss of phosphorus from peat–sand mixes by leaching.

Fertilizer	% phosphorus lost in first leaching	Total % phosphorus lost in seven leachings
superphosphate	56.2	84.5
MagAmp	7.8	73.0
potassium metaphosphate	0.1	63.5
Frit 9144	2.2	12.6
Frit 46	21.3	54.8
Enmag	0.7	40.8
basic slag	3.8	31.8
bonemeal	1.1	3.1
Gafsa rock phosphate	0.2	0.5

in Figure 6.2. Ground rock phosphate and bonemeal had very low availabilities and for most purposes would not be considered suitable sources of phosphorus. Potassium metaphosphate, magnesium-ammonium-phosphate, basic slag and Frit 9144 all showed useful slow-release patterns; they could be considered as 'reserve' sources of phosphorus in mixes if some superphosphate was also added for immediate use by the plant.

The availability of phosphatic fertilizers is influenced by granule size as well as its chemical form. Results from an unpublished experiment by the writer with granular and powder forms of fertilizers are given in Table 6.3; the experiment was made on similar lines to the previous one. Granules of each fertilizer were ground to provide the powder form, and the results have been expressed relative to those with superphosphate. As well as having a low loss by leaching, the ideal fertilizer must also be available to plants; the Gafsa phosphate can therefore be rejected because of its low availability. With the other fertilizers the leaching loss in the granular

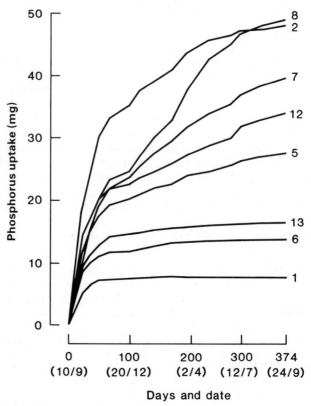

Figure 6.2 Availability of various forms of slow-release phosphorus sources to rye grass (*Lolium perenne* cv. S24).
Key:1 control; 2 superphosphate; 5 Frit 9144; 6 Gafra rock phosphate; 7 Enmag; 8 potassium metaphosphate; 12 basic slag; 13 bonemeal.

128

Table 6.3 Comparison of superphosphate and slow-release phosphorus fertilizers in powder and granular forms. Resistance to leaching and plant uptake expressed as the percentage of superphosphate values (P = powder form, G = granular form).

Fertilizer	Leaching loss		Plant uptake	
	first two leachings	six leachings	after 87 days	after 304 days
superphosphate	100	100	100	100
triple superphosphate (P)	103	104	102	104
triple superphosphate (G)	64	77	103	109
dicalcium phosphate dihydrate (P)	77	81	105	103
dicalcium phosphate dihydrate (G)	18	27	68	52
magnesium-ammonium-phosphate (P)	78	84	106	102
magnesium-ammonium-phosphate (G)	22	49	71	75

	Particle size analysis of fertilizers (% weight)		
	>3 mm	3–2 mm	2–1 mm
triple superphosphate	33.73	60.96	5.29
dicalcium phosphate	34.93	45.72	16.93
magnesium-ammonium-phosphate	3.31	43.84	50.98

form was significantly lower than the powder form. Granular phosphate dihydrate had a very low loss, and even the granular form of triple superphosphate had a very useful initial resistance to leaching. Granules of triple superphosphate dissolve much quicker than either magnesium-ammonium-phosphate or dicalcium phosphate dihydrate. In powder form all these fertilizers had the same availability to the grass as superphosphate. The order of availability in granular form was: triple superphosphate ($\equiv$ powdered single superphosphate) > magnesium-ammonium-phosphate > dicalcium phosphate dihydrate. Granular forms of the two latter materials have been the best of slow-release phosphorus fertilizers tested for use in peat and lightweight mixes.

6.1.4 *Phosphorus toxicity*

Whereas phosphorus is required in much smaller amounts than either nitrogen or potassium, most plants are able to tolerate relatively high levels in the growing media and only show distress when it interacts to affect the availability of other elements, e.g. an induced iron deficiency. However, the very wide range of plants now being grown in containers, coupled with the use of organic-based potting mixes which do not 'fix' phosphorus in the same way as do soil-based potting mixes, have resulted in the occurrence of phosphorus toxicity with some species.

Plants such as tomato and *Chrysanthemum* are naturally tolerant of high phosphorus levels, also any susceptibility to phosphorus toxicity would have been eliminated during the extensive breeding and selections made over many generations. Other plants however, that have evolved by natural selection in soils of low fertility, have shown phosphorus toxicity when grown in container mixes. Two such groups are the Australian and South African *Proteaceae* and some of the hardy ornamental flowering shrubs, e.g. *Cytisus* x *praecox*. Thomas (1980) grew a range of plants in a mixture of equal parts of sphagnum peat and perlite to which nitrogen, potassium and varying rates of superphosphate were added. It was concluded that for many of the proteas the rates of superphosphate normally used in soil-less mixes are too high. Some of these plants, e.g. *Hakea laurina*, are intolerant of orthophosphates at rates above 50 mg l^{-1} phosphorus whereas for others, e.g. *Grevillea rosmarinifolia*, the maximum safe rate of phosphorus is 100 mg l^{-1}. The proteas have adapted to soils very low in phosphates by producing proteoid roots which can store phosphorus as polyphosphates. Other plants reported as being susceptible to high levels of water-soluble phosphorus are subterranean clover, oats and some species of lupins. The symptoms include blackening of the shoot tips, the lower leaves become pale, necrotic and are then shed from the plant. Evidence of an interaction between salinity and the level of orthophosphates in the medium has been reported by Niemanand & Clark (1976). Salinity appears to damage plant mechanisms that normally regulate the amount of orthophosphates in the tissue, resulting in excessive accumulation and phosphorus toxicity.

6.1.5   Fluoride toxicity

Although phosphatic fertilizers have a higher content of microelements than most nitrogeneous and potassium fertilizers, the amount of microelements supplied in this way is low and is often regarded as beneficial rather than detrimental. An exception, however, is the amount of fluorides present in superphosphate; these can be toxic to some plants. Conover & Poole (1982b) have classified a number of foliage plants on their susceptibility to fluoride. Those considered most sensitive to injury were: *Cordyline terminalis* 'Baby Doll' and *Dracaena deremensis* 'Janet Craig'; plants moderately affected were *Chlorophytum comosum* and *Yucca elephantipes*. They tested numerous materials used in potting mixes for soluble fluorides; values over 100 p.p.m. dry weight were considered high and included superphosphate (2600 p.p.m.), diammonium phosphate (2000 p.p.m.) and triple superphosphate (1600 p.p.m.). For sensitive plants, fluoride toxicity can be avoided by: raising the pH of the mix to 6.0–6.5, using phosphoric acid instead of superphosphate and avoiding irrigating with water having more than 0.25 p.p.m. of fluoride. Other crops in which fluoride toxicity has been reported are palms, *Gladiolus*, *Lilium* and *Freesia*.

6.1.6 Mycorrhiza

Mycorrhizal fungi form symbiotic associations with the roots of many plants. Those of greatest interest in the production of woody plants in containers are known as vesicular-arbuscular mycorrhizae (VAM). The fungal hyphae grow from the roots into the soil for several centimetres and absorb nutrients, especially phosphorus, which are transferred to the plant. Container mixes usually have high levels of available phosphorus and the response of most woody plants to VAM inoculation is therefore small. In other instances, however, plant growth has remained stunted in the absence of VAM infection even when heavy phosphorus fertilization has been given, indicating a mycorrhizal dependency, e.g. *Magnolia grandiflora* and citrus. Other examples of woody plants which respond to VAM inoculation are the sweet gum and yellow polar. Some of the beneficial effects of VAM are believed to occur when plants have coarse root systems with very few root hairs. On transplanting, container plants grown with VAM inoculation should become better established in soils with very low levels of available phosphorus than would non-inoculated plants.

6.2 POTASSIUM

The amount of potassium present in the tissue of pot plants is often of the same magnitude as that of nitrogen, an average value being 3–4% of the dry weight. However, when plants are grown with high levels of potassium in the mix, the amount of potassium present in the plant tissue can be very high: this condition is often referred to as 'luxury consumption'.

Potassium originates from such minerals as the micas and felspars, and an average value for potassium in a mineral soil would be about 2%. The amount of potassium present in organic materials is usually much lower than that in mineral soils. An average value for peat would be about 0.04% of the oven-dry weight, and when an allowance has been made for the very much lower bulk density (about 0.1–0.2 g cm^{-3} for peat, by comparison with 1.2 g cm^{-3} for a mineral soil), it will be readily seen that loamless mixes based upon peat will have a very low natural content of potassium. Vermiculite is the only material used in making loamless mixes which contains significant amounts of potassium. In mixes made with materials having a base exchange capacity, the potassium will be present in both the water-soluble and exchangeable forms. As the potassium present in the soil solution is either absorbed by the plant or lost by leaching, some of the exchangeable form is released to take its place. Similarly, when water-soluble potassium is applied in the liquid feed, some of it will be taken into the exchangeable form.

Potassium is necessary for the function of enzymes that control carbohydrate and nitrogen metabolism. It also acts as an osmotic regulator in the water relations of plants; there is some evidence that the plant can partially substitute sodium for potassium. Horticulturists normally associate it with balancing the soft growth resulting from too much nitrogen; such plants are generally considered as being more susceptible to fungal diseases. It is also essential for the production of high quality tomato fruit, where large quantities of potassium are supplied in relation to the amount of nitrogen. Often the liquid fertilizers applied to tomato plants have a ratio of 1 nitrogen:2 potassium, whereas for azalea culture the reverse ratio of 2 nitrogen:1 potassium would be more usual.

Deficiency symptoms in plants often start with mottling of the foliage and pale leaf margins. As the deficiency increases, necrotic spots develop, the leaf edges scorch and turn brown, and the lower leaves may absciss.

6.2.1   Potassium fertilizers

Potassium does not have such complex reactions within the media as either nitrogen or phosphorus. Only relatively few chemicals are suitable for use as potassium-supplying fertilizers and, with few exceptions, they do not present any problems.

Potassium sulphate (K_2SO_4)   This fertilizer contains 43% potassium and 18% sulphur; it is a dry powder which does not readily absorb water vapour from the atmosphere and is easy to handle. Although the commercial grade of fertilizer supplies potassium in a water-soluble form, the impurities present prevent it being used to prepare the highly concentrated liquid fertilizer stock solutions which are diluted before being applied to the plants. The commercial grade fertilizer can be used only as a base fertilizer in mixes; a more refined grade containing fewer impurities and having a higher solubility must be used if it is desired to use potassium sulphate in the liquid feed.

Potassium chloride (KCl)   Also known as muriate of potash, this material contains 51% potassium and has a high degree of water solubility. It contains 47% chloride and, because of the risk of chloride toxicity, is not generally used in Britain for glasshouse crops; however, it has been the most widely used potassic fertilizer in the USA.

Potassium carbonate (K_2CO_3)   This material contains 55% potassium; it is basic in its reaction and can be used where it is desired to raise the pH of the mix. It is not in common use in Britain.

Potassium nitrate (KNO_3)   In addition to supplying 38% potassium, this fertilizer also contains 13% of nitrate nitrogen. It has a high degree of water solubility which makes it the obvious choice of a potassic fertilizer for preparing liquid feeds, provided that the amount and form of nitrogen is acceptable. It can also be used in the base fertilizer but is slightly hygroscopic and for this reason must be stored under dry conditions.

Potassium frit   The term 'frit' or 'fritted' refers to the process of fusing a potassium salt with molten sodium silicate. This is rapidly cooled by pour-ing it either onto a cold surface or into running water where it solidifies and shatters into small pieces which are then finely ground. This gives a material that has 29% potassium in a form which is not immediately water soluble, but which dissolves slowly in the medium and releases potassium over a long period. This form of slow-release action is not usually consi-dered necessary in the case of potassium, where the initial concentration in the mix can be readily controlled by regulating the strength of the base fertilizer. The potassium level can then be easily maintained by adjusting the strength of the liquid fertilizer. The fritted type of fertilizer has a much greater potential for supplying micro- or trace elements (Ch. 7).

6.3   CALCIUM

This element is required by the plant in regions of active cell division for the formation of cell walls in which it occurs as calcium pectate. It also plays a role in the transport of carbohydrates and amino acids within the plant, and in the development of new roots.

A deficiency of calcium shows as a stunting of plant growth. Leaves are restricted in their development and can show a paleness at the margins, which may be followed by an inward curling. These leaf symptoms, which are most noticeable in broad-leafed plants, may be followed by necrotic spotting. In cases of acute deficiency, the tips of the plants will die and lateral buds remain dormant. Roots are particularly sensitive to calcium deficiency; young roots and root hairs are killed and the older roots turn brown. Wind (1968) found that very little root growth occurred in peat when the pH was below 3.5 (measured in 0.1 M KCl). Gammon (1957) studied the growth of roots in soils adjusted to different pH values by the use of carbonates of calcium, sodium or potassium, and concluded that the lack of calcium was more detrimental to root growth than the pH itself. Paul & Smith (1966) reported a sharp deterioration in the rooting of chrysanthemum cuttings in peat when the exchangeable calcium was less than 37.5% of the total exchange capacity (pH 4.4). Below this value the root tips were brown and both the total amount of root per cutting and the mean root length declined. It was concluded that calcium saturation should be between 37.5 and 75%. In another study with five woody species, the optimal percentage of exchangeable calcium for rooting varied with the species; in general, root formation was best when the exchangeable calcium was 40% (pH 5.0) or more (Paul & Leiser 1968).

Calcium is also required for fruit and seed development, a good example of calcium deficiency being blossom-end rot in tomatoes. This is usually seen at the distal or blossom end of the fruit as a water-soaked region, which develops into blackened, hard, dry areas; symptoms may also be present within the fruit but not visible from the outside. This condition can be induced by high salt stress, dryness, and high levels of soluble potassium, magnesium or ammonium, all of which reduce the uptake of calcium by the plant.

Apart from their effects upon the soil salinity and certain antagonistic effects towards one another, most elements can be regarded as having only one prime function, i.e. their direct role in plant nutrition. Calcium, however, also has an important indirect role. It is the element normally found in the greatest concentration in the mix and has a greater effect in controlling the reaction or pH of the mix than any other element. In addition to its removal in relatively large quantities by the plant, usually of the same order of magnitude as for nitrogen and potassium, it is also lost in large amounts by leaching. The fertilizer, ammonium sulphate, is generally regarded as causing a loss of calcium carbonate equal to its own weight, and this is followed by a drop in the pH. The quantity of calcium required to change the pH is related to the buffer capacity and cation exchange capacity of the medium (see § 4.7).

Calcium is therefore important in plant nutrition because of:

(a) its function within the plant;
(b) its effect on the pH and thereby on the availability of other macro- and microelements;
(c) its effect on disease control, e.g. liming to a high pH helps to control clubroot in brassicas and *Fusarium* in tomato plants, whereas a pH of not greater than 5.0 is considered helpful in controlling *Thielaviopsis* root rot in *Poinsettia*.

6.3.1   Calcium-supplying materials

The most common forms of calcium or liming materials are calcium oxide, calcium hydroxide and calcium carbonate.

Calcium oxide (CaO)   Commonly known as burnt lime, quicklime or unslaked lime, it is obtained by heating calcium carbonate and driving off the carbon dioxide. This gives a highly concentrated form of lime which can be caustic to the roots of young seedlings; it can also be disagreeable to handle. It is often used in agriculture but is not recommended for use in potting mixes.

Calcium hydroxide (Ca(OH)$_2$)   Also known as slaked lime, quenched lime or hydrated lime, this form has a solubility of 1700 p.p.m. in water and can also have a caustic action on young seedlings. If used as a liming material in mixes, an interval of several days should elapse between mixing and using the media. Its appeal lies in achieving a more rapid increase in the pH than is obtained with chalk; also, being in a more concentrated form than chalk, less material is required. Both calcium oxide and calcium hydroxide eventually revert to calcium carbonate in the media.

Calcium carbonate (CaCO$_3$)   This material is also known in horticultural circles as chalk, carbonate of lime and ground limestone. It is an inherently safe form of lime with no caustic action, and is the form recommended for

Table 6.4   The relative neutralizing values of various liming materials.

Liming material	Molecular weight	Neutralizing value	
		$CaCO_3 = 100\%$	$CaO = 100\%$
calcium oxide (CaO)	56	178	100
calcium hydroxide (Ca(OH)$_2$)	74	135	75
calcium carbonate (CaCO$_3$)	100	100	56
magnesium oxide (MgO)	40	250	140
magnesium hydroxide (Mg(OH)$_2$)	58	172	96
magnesium carbonate (MgCO$_3$)	84	119	67
Dolomite limestone	—	108	60

use in potting mixes. Its solubility in pure water is only 10 p.p.m. but this increases to 600 p.p.m. when the water contains carbon dioxide in solution. Its effectiveness or speed of action in the mix is largely determined by its particle size; the finer the particles, the more rapid its effect. For use in potting mixes not less than 90% by weight should pass a 100-mesh sieve and 50% a 200-mesh sieve.

The efficiency of these materials for liming purposes can be compared from their molecular weights, and this is often referred to as their neutralizing value; unfortunately, some countries adopt calcium carbonate as the reference value, and others use calcium oxide. The comparative neutralizing values of various liming materials are shown in Table 6.4. Other materials, such as wood ash and basic slag, also have some neutralizing value, but they are not standardized materials and are not to be recommended.

Calcium sulphate (CaSO$_4$)   Commonly known as gypsum, this can be used whenever it is desired to add calcium to the medium without appreciably affecting the pH value; only under very acid conditions will calcium sulphate raise the pH.

6.4   MAGNESIUM

Magnesium is one of the elements required for the formation of chlorophyll and a deficiency will soon lead to a reduction in the rate of photosynthesis. It also has an effect on the transport of phosphorus within the plant.

Commonly observed deficiency symptoms are a yellow-green mottle, usually appearing first on the older leaves, the veins remaining green. In some plants a deep red pigment develops in the centre of the leaves and this may be followed by necrotic spots; this symptom is frequently seen in pot-grown tomato plants. Magnesium deficiency may be due to low levels of exchangeable magnesium in the mix but often it is the result of very high levels of potassium rather than a true deficiency of magnesium.

Branson *et al.* (1968) investigated this problem in chrysanthemums and found that, at an early stage in the life of the crop, the leaves had 0.1% magnesium and 4.8% potassium (dry weight basis); as the crop developed the amount of magnesium in the leaves fell to 0.04% and deficiency symptoms appeared. Adding magnesium sulphate to the soil was not found to be very effective in reducing the deficiency unless the feeding of potassium was discontinued.

Experiments at the Glasshouse Crops Research Institute have shown that magnesium deficiency symptoms are unlikely to occur when the K:Mg ratio in the mix is 3:1 or less. Hooper (personal communication) has also concluded from the examination of large numbers of mixes for advisory purposes that plants are free from deficiency symptoms when the ratio of extractable K:Mg is below 3:1. At a ratio of 4:1 and above, the risk of deficiency symptoms was found to be much greater.

Magnesium deficiency in crops such as tomato and chrysanthemum can also be controlled by applying a foliar spray; usually a 2% solution of Epsom salts is used. There is some evidence that the rate of magnesium absorption by the leaves is faster with magnesium nitrate and magnesium chloride than with magnesium sulphate. The leaves of some plants can be scorched if solutions stronger than 2% are used.

In water culture experiments, magnesium sulphate and magnesium chloride have been more toxic to plants than the sodium and calcium salts at similar osmotic concentrations, indicating a specific toxicity of magnesium. Paul & Thornhill (1969) rooted chrysanthemum cuttings in peats having differing ratios of exchangeable calcium and magnesium; when the amount of exchangeable magnesium was more than 80%, rooting was severely impaired. The rooting response also dropped dramatically when the amount of magnesium in the mist water exceeded 60% of the total cations.

6.4.1 Magnesium-supplying fertilizers

Although several magnesium supplying materials are available, normally only two forms are in common use in potting mixes.

Magnesium sulphate This is available as Epsom salts ($MgSO_4 \cdot 7H_2O$), with 10% magnesium, and as Kieserite ($MgSO_4 \cdot H_2O$) which has less water of crystallization and therefore a higher magnesium content (16%). Kieserite is also cheaper than Epsom salts. Magnesium sulphate is sometimes used at rates up to 1 kg m^{-3} (1.6 lb yd^{-3}) in potting mixes; however, it is usually safer to use magnesium limestone.

Magnesium limestone This material is a calcium-magnesium carbonate containing 11–13% magnesium; a low grade material having only 3% magnesium is also available. The terms Dolomite limestone and Dolomitic limestone are often incorrectly used synonymously with magnesium limestone. In correct usage, Dolomite refers to a compound containing equal amounts of calcium and magnesium carbonate with 12%

magnesium. When the limestone contains very little magnesium it is termed *calcic*, as the magnesium content of the material rises it is called *Dolomitic*, and when the material is composed almost entirely of calcium-magnesium carbonate it is then called *Dolomite*. Thus, in practice, Dolomitic limestone has a lower magnesium content than Dolomite limestone.

In addition to its function of supplying magnesium, this material is also used to neutralize soil acidity. By laboratory methods, it usually shows approximately the same neutralizing value as calcium carbonate, but, under practical conditions, it is somewhat slower than calcium carbonate in raising the pH, as magnesium carbonate is less easily dissolved by mineral acids than is calcium carbonate. In some situations, however, magnesium carbonate is preferable as a liming material to calcium carbonate. Tod (1956) grew *Rhododendron davidsonianum* in a soil to which large amounts of magnesium carbonate were added. This raised the pH to levels which would have been considered toxic to rhododendrons if calcium carbonate had been used. However, no serious ill effects resulted from its use and it was concluded that alkalinity *per se* was not the cause of the harmful effects produced by alkaline soils on rhododendrons.

Magnesium limestone is also available in the burnt form; the calcium and magnesium are then present as oxides. However, magnesium oxide is less soluble than calcium oxide and it takes up carbon dioxide less readily, consequently it remains in a caustic condition in the mix longer than calcium oxide. For this reason burnt magnesium limestone is not recommended for use in potting mixes.

Experiments made at the Glasshouse Crops Research Institute have shown that large interactions occur between the forms of the liming materials and the sources of nitrogen (Fig. 6.3). Using urea as the nitrogen source, Dolomite limestone was superior to calcium carbonate but was inferior to calcium sulphate. With calcium nitrate as the nitrogen source, plant growth was similar with all three liming materials. The toxic effect of the urea was related to the pH of the mix; the higher pH obtained with the calcium carbonate caused the production of free ammonia and nitrate nitrogen. Toxicities due to these forms of nitrogen have been discussed in Chapter 5. Dolomite limestone, used at the same rate as calcium carbonate, gave a slightly lower pH and the growth depression was correspondingly reduced; calcium sulphate had the least effect upon the pH and gave the best growth. These results illustrate the importance of achieving the correct balance in plant nutrition and also the need to exercise caution in changing the formulation of a mix. Simply substituting one form of nitrogen or liming material for another can often have significant effects on plant growth, unless the possible interactions of the fertilizers are first considered.

Peats differ from mineral soils in having a much greater percentage of magnesium present in an exchangeable form, and, providing that some magnesium has been included in the base fertilizer, deficiency symptoms are not liable to occur, unless the potassium level is allowed to rise excessively. One plant which is very susceptible to magnesium deficiency

Figure 6.3   The interaction of nitrogen source and liming material. *Top*: urea as the nitrogen source. *Bottom*: calcium nitrate as the nitrogen source. *Left*: calcium carbonate. *Centre*: Dolomite limestone. *Right*: calcium sulphate.

induced by high levels of potassium is the winter cherry (*Solanum capsicastrum*). Owen (1948) investigated this problem with regard to potting composts based on a mineral soil. The most effective treatment was the addition of Epsom salts (magnesium sulphate) in the base fertilizer at rates equal to the amount of potassium sulphate used. If the potassium had been supplied by a compound or complete N P K fertilizer, then the amount of magnesium sulphate should be equal to twice the amount of K_2O in the compound fertilizer.

6.5   SULPHUR

This element is an important part of the amino acids or nitrogenous compounds from which proteins are formed. Plants are unable to utilize sulphur in the elemental form; it must be absorbed by the roots as sulphates. Sulphur deficiency symptoms are seldom seen in pot-grown plants which are normally adequately supplied with sulphates from fertilizers such as superphosphate, ammonium sulphate, etc., and also as an impurity in the irrigation water; an excess of sulphate is usually a greater problem than a deficiency. Fortunately, plants are more tolerant of an excess of sulphate than they are of chloride. Eaton (1942) found that chlorides were approximately twice as toxic as sulphates to the tomato, bean and lemon. Saturated solutions of calcium sulphate have been applied to *Primula*, carnation, *Chrysanthemum*, *Fuchsia* and *Poinsettia* without any harmful effect; azaleas and Rex begonias, however, have shown unfavourable reactions (Pearson 1949). High concentrations of sulphates in the mix present problems when salinity determinations are made using wide-ratio suspensions rather than saturated paste extracts. The precautions necessary in making salinity determinations under these conditions have been discussed in sections 4.8.3 and 4.8.4.

By comparison with mineral soils, some peats have a relatively high sulphur content which oxidizes to sulphates when they are drained and cultivated, deposits of calcium sulphate can sometimes be seen in the drainage ditches of bogs.

6.6   SLOW-RELEASE FERTILIZERS

Fertilizers used in potting media preparation can be classified into two groups: (a) water-soluble and (b) non water-soluble. Those in the former group are immediately available to plants; their disadvantages include raising the salinity of the media with possible retardation of growth or plant injury, and low retention by organic and lightweight mixes, thereby necessitating early and frequent supplementary feeding.

Fertilizers that are not immediately water-soluble can be loosely classified into three types:

Type	Example	Principal factors controlling availability in the mix
organic	hoof and horn, urea-formaldehyde	bacterial and fungal activity
low solubility	dicalcium phosphate, magnesium-ammonium-phosphate	particle size
controlled release fertilizers	Osmocote, Nutricote	temperature

The organic fertilizers are principally suppliers of nitrogen, their general characteristics and use in potting mixes have been discussed in Chapter 5.

The low solubility type fertilizers that are of most interest for use in potting mixes are the granular forms of magnesium-ammonium-phosphates and dicalcium phosphate. Although the former supply small amounts of nitrogen and sometimes potassium, they should be regarded as slow-release phosphorus fertilizers and have been included under that section (§ 6.1.2) as well as in Chapter 5.

139

6.6.1   Controlled release fertilizers

Characteristics   A wide range of materials have been used to coat prills of inorganic fertilizers to reduce their immediate solubility and availability to plants; materials used for coating include acrylic resins, polyethylene, waxes, latex and sulphur. Resin- and polymer-coated fertilizers are the most widely used in potting mixes, they include:

Osmocote, manufactured by Sierra Chemical Co., Milpitas, California, USA. The main constituent of the coating is a copolymer of dicyclopentadine with glycerol ester (Powell 1968).

Nutricote, manufactured by Chisso-Asaki Fertiliser Co., Tokyo, Japan. The coating consists of copolymers of polyolefinic and polyvinylidene chloride (Shibata *et al.* 1980).

The nutrients are released by the action of water vapour moving into the capsule through the very small pores in the coating; the dissolved salts then diffuse into the mix. The rate of release is mainly dependent upon the temperature of the media, the thickness of the coating and the number of pores it contains. One other factor which affects the release is the chemical formulation of the compound in the prill; salts such as ammonium nitrate absorb water vapour much more quickly than do potassium nitrate or potassium sulphate. This will affect not only the rate of solution, and therefore the release rate, but also the relative speeds at which the different elements are released. By varying the formulation of the prill and the type of coating it is possible to make fertilizers with a wide range of NPK analysis and release rates.

Examples of the formulations available ($N : P_2O_5 : K_2O$ %) are:

	USA	UK
Osmocote	14–14–14, 3–4 months 18–6–12, 8–9 months	15–11–13 + 2MgO, 3–4 months 18–11–10, 8–9 months
Ficote		14–14–14, 70 days 16–10–10, 140 days
Nutricote (both USA and UK)	13–13–11, 100 days 16–10–10, 140 days	

Formulations used in different countries may vary slightly because of different salts in the prill or because the type of crop and management require a different analysis. Another type of fertilizer available in the USA and the UK is known as 'Precise'; this is a liquid fertilizer within a spherical semi-permeable capsule. It is mainly used by hobby and amateur gardeners.

Recommended rates of use for controlled release fertilizers vary with the crop, the local environment and management. Typical rates would be up to 4 kg m^{-3} (6.5 lb yd^{-3}) of the 3–4 month formulation for pot chrysanthemums grown without liquid feeding, and 3 kg m^{-3} of the 8–9 month formulation for hardy nursery stock. Assuming that the maximum rate of soluble nitrogen that can safely be added to a potting mix is about 250 mg l^{-1}, the above rates of slow-release fertilizers are supplying nitrogen at $2\frac{1}{4}$ times that rate.

Once the coated fertilizer has been mixed with the medium and it is watered, the main factor controlling the release rate is temperature (Oertli & Lunt 1962); pH, water content and bacterial activity have no significant effects. Release rates are usually determined by immersion in water at a constant temperature and are quoted as the number of days required to obtain a release of 80% of the fertilizer. However, Shibata *et al.* (1980) found that the release rate in water was between 1.2 and 1.5 times faster than in soil; also the relative release rates of N:P$_2$O$_5$:K$_2$O for Nutricote was 1:0.6:0.9. This confirms early observations that with this type of fertilizer and some crops it is desirable to add superphosphate when preparing the mix. The extent to which temperature and type of coating determine the release rate for Osmocote is clearly seen in the results of Harbaugh & Wilfret (1982), Figure 6.4. For the 14–14–14 formulation, release of 50% of the salts occurred in 15 days at 86 °F (30 °C), in 26 days at 73 °F (23 °C) and in 50 days at 60 °F (16 °C); for the 18–6–12 fertilizer in the 9–12 month formulation it required 96 days at 73 °F (23 °C) for 50% release. It should also be noted that with the 3–4 month formulation 20–25% of the final salt value was obtained within the first three days; this early release is due to imperfections within the coating. Worrall (1981b) compared the release rates of Osmocote and Nutricote at a range of temperatures. For the 8–9 month formulations Osmocote released 53% after 10 weeks at 35 °C and Nutricote 31%; this agrees with grower experience of the two fertilizers. At temperatures above 35 °C the release rates were much higher, especially with Osmocote. The release patterns published by various workers for coated and slow-release fertilizers often show poor agreement between results obtained in the laboratory and those based on plant growth. This is sometimes due to artifacts introduced into the laboratory methods, and also to interactions between the fertilizer and the system of plant management.

Controlled-release fertilizers (CRF) or liquid feeding Detailed examination of the growth rates of young seedlings in media to which varying amounts of water-soluble fertilizers have been added will soon show that the first response to even a moderate amount of fertilizer will be a depression of the growth rate (see Fig. 8.2). Conversely, the low fertilizer level that gives the best growth rate for young seedlings is unable to sustain growth for very long before nutrients become limiting. The amount of soluble fertilizer in the mix must therefore always be a compromise between these two situations. In theory, controlled-release fertilizers (CRF) offer several advantages:

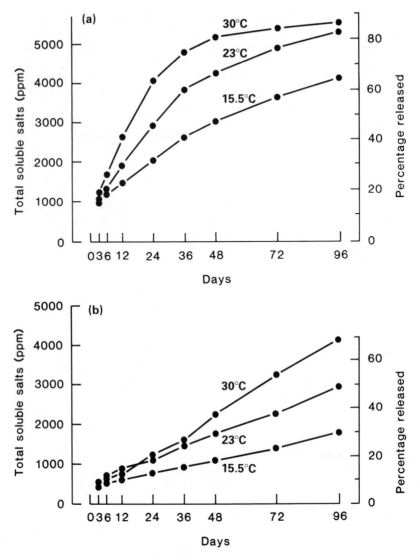

Figure 6.4 Effect of temperature on the release rates of coated fertilizers, (a) Osmocote 14–14–14, 3–4 month formulation; (b) Osmocote 18–6–12, 9–12 month formulation.

(a) avoidance of the high initial salt levels;
(b) nutrients are released progressively as plants grow;
(c) possible reduction in the amount of nutrients lost from the container by leaching;
(d) liquid feeding requires expensive equipment and frequent monitoring, and in some situations, e.g. in nursery stock, overhead feeding can be very wasteful with only a small amount of fertilizer actually entering the container.

The amounts of nutrients that must be given to plants in containers to maintain their maximum growth rate is principally determined by the environment, i.e. the potential for growth, and the quantity of nutrients that are lost by leaching. Precise values cannot be given for either of these factors, and neither will the rates remain constant over the life of the plant. As an example of the way in which the environment affects the demand for nitrogen, pot chrysanthemums grown in southern England in summer assimilate about three times as much nitrogen as in winter.

The results of Hershey & Paul (1982) provide the best assessment of the amounts of nitrogen lost by leaching when chrysanthemums are grown with either CRF or liquid feeding. With CRF, the nitrogen in the leachate declined from an initially high value to almost zero at six weeks, and then remained very low for the following four weeks. Liquid feeding gave a completely different pattern, the loss increased progressively over the ten weeks. Comparison of nitrogen losses shows that on the basis of *equal amounts of nitrogen being applied* by the two methods, leaching losses with liquid feeding were about twice those with the CRF. However, when comparison was made on the basis of *the amount of nitrogen that had been released from the CRF* the difference between the two systems was much less, but still in favour of a lower loss by CRF than by liquid feeding. Losses with either system will of course be dependent upon the leaching factor, i.e. the amount of excess water or liquid feed applied to the container. Evidence from other experiments shows that although it is difficult to judge the correct amounts of CRF required to maintain the maximum growth rate, leaching losses will usually be greater with a system based on liquid feeding. Frequently the best growth is obtained where a moderate application of CRF is supplemented with some liquid feeding. When seedlings are transplanted into a mix containing CRF it will be an advantage to delay liquid feeding for up to seven waterings to avoid high salinities (Johnson 1973a).

6.7   MINERAL SOIL AND PEAT COMPARISON

The quantity of each of the macroelements present in the various types of peats and mineral soils in their natural states can be very variable, being dependent upon the basic types of vegetation and minerals from which the peats and the soils have been formed. Simplified comparisons of the mineral values of these two media must, therefore, be treated with some caution and the values given in Table 6.5 should be regarded only as showing the large differences that can exist between the two types of media. It must also be noted that the values in the table are on a weight basis, and the bulk density of peat is only one tenth or less of that of a mineral soil. When the nutrient values are transformed to a volume basis, those of peat will be reduced correspondingly. Apart from the sulphur and the organic matter, peats will be seen to have a much lower content of plant nutrients than mineral soils. We have already noted, however, that when fertilizers are added to the two media at equal rates per

Table 6.5 Comparative mineral values of an average peat and an average mineral soil, % by weight. (The bulk density of peat is only approximately one tenth that of mineral soil.)

Nutrient	Peat	Mineral soil
organic matter	95.0	3.0
C:N ratio	30:1	10:1
nitrogen	2.5	0.15
phosphorus	0.03	0.05
potassium	0.04	1.80
calcium	0.20	0.35
magnesium	0.15	0.30
sulphur	0.15	0.04

unit volume, the levels of available plant nutrients in the peat-based mixes can be expected to be higher than those of mineral soil composts. This is because of the lower amount of 'fixation' that occurs in peat-based mixes.

6.8 NUTRIENT AND ENVIRONMENT INTERACTIONS

In studies at the Glasshouse Crops Research Institute on the nutrition of plants grown in peat-based mixes, several interactions have been observed between certain macroelements; interactions have also occurred between the nutrition and the environment or the season.

An example of an interaction between macronutrients is shown in Figure 6.5. Antirrhinums were grown in peat–sand mixes having either calcium nitrate or hoof and horn as the nitrogen source; with each fertilizer there were two rates of calcium carbonate application. At the high rate (pH 6.0 in 0.01 M $CaCl_2$), only calcium nitrate was an acceptable

Figure 6.5 The interaction of nitrogen source and rate of calcium carbonate. *Left to right*: calcium nitrate, high rate of calcium carbonate; hoof and horn, high rate of calcium carbonate; calcium nitrate, low rate of calcium carbonate; hoof and horn, low rate of calcium carbonate.

144

Figure 6.6 The interaction of the rate of organic nitrogen, the rate of calcium carbonate and the season. *Top illustration*: tomato plants grown in four mixes from a factorial experiment in winter. *Top left*: low nitrogen, low calcium carbonate. *Top right*: high nitrogen, low calcium carbonate. *Bottom left*: low nitrogen, high calcium carbonate. *Bottom right*: high nitrogen, high calcium carbonate. *Lower illustration*: the same treatments, grown in summer.

source of nitrogen; the hoof and horn treatment showed marked phyto-toxicity and growth depression. At the lower rate (pH 4.9), there was no significant difference in growth between plants in the two sources of nitrogen.

An example of the type of interaction which can occur between the nutrition and the environment, where tomato plants were grown in two factorial design experiments, is seen in Figure 6.6. The four treatments had received calcium carbonate and nitrogen at either low or high rates of application, the nitrogen being 70% in the form of hoof and horn, and 30% as ammonium nitrate. Other nutrients were given at constant rates for all the treatments. In winter, there were large differences in growth between the four treatments. Plants having the high rate of nitrogen and the high rate of calcium carbonate application made the least growth; often this treatment resulted in the eventual death of the plants. However, with plants at a comparable stage of development in summer there was very little difference in growth between the four treatments, and eventually the high nitrogen treatments produced the most growth. The beneficial effect of the high rate of superphosphate (see Fig. 6.1) is also much greater in winter than it is in summer. However the requirement of phosphorus by the plants is actually lower in winter than it is in summer, because of their lower growth rate under the low light conditions of winter. Some of the reasons for this beneficial effect of superphosphate have been discussed in section 6.1.1. Because of the higher media temperatures, the rates of mineralization of hoof and horn and some other forms of slow-release fertilizers are approximately twice as high in summer as in winter. The rates of plant growth, however, can be ten or more times greater in summer than winter; consequently there is less risk of nitrogen toxicity. True mineral deficiencies are more quickly observed in summer than in winter.

6.9   FERTILIZER ANALYSIS AND SALT INDEX

A summary of the most commonly used fertilizers and their major plant nutrient contents is given in Table 6.6. The nutrient content may vary slightly in different countries, depending upon the precise method of fertilizer manufacture and also the sources and purity of the basic materials used.

All soluble fertilizers contribute towards the salinity of the soil solution, some more than others. As already seen (Ch. 4), high salinity values are detrimental to plant growth and the aim should be to choose a fertilizer which, for a given amount of plant nutrients, has the least effect on soil salinity. Rader *et al.* (1943) determined the effects of *equal weights of fertilizers* upon the osmotic potential of the soil solution, these data are given in Table 6.7. To enable the more useful comparisons to be made of the effect of *equal amounts of nutrients*, which is the basis used in formulating potting mixes, the data have been recalculated to give the salt index of

Table 6.6 Nutrient analysis of the fertilizers commonly used in preparing loamless mixes.

Fertilizer	Analysis				
	nitrogen	phosphorus	potassium	calcium	magnesium
ammonium nitrate	35.0				
ammonium sulphate	20.5				
calcium nitrate*	15.5			19	
sodium nitrate	16.0				
urea†	46.0				
basic slag		3–9		32	
calcium metaphosphate		27.5		17.8	
diammonium phosphate	21	23			
monoammonium phosphate	12	26.6			
superphosphate (single)		8–10		17–22	
superphosphate (triple)		19			
ammonium polyphosphate	15	27			
phosphoric acid		24			
superphosphoric acid		33			
dicalcium phosphate		18		23	
potassium carbonate			55		
potassium chloride			51		
potassium dihydrogen phosphate		22.8	28.7		
potassium metaphosphate		25	32		
potassium nitrate	13		38		
potassium sulphate			43		
calcium carbonate				40	
calcium hydroxide				54	
calcium oxide				71	
calcium sulphate				57	
magnesium carbonate					11–13
magnesium hydroxide					41
magnesium oxide					60
magnesium sulphate (Epsom salts)					10
Kieserite					16

* A more pure form that does not contain any ammonium nitrate has an analysis of 12% nitrogen and 17% calcium.

† The use of urea in mixes is not recommended.

Table 6.7 Effect of fertilizers on the soil salinity.

Fertilizer		Salt* index	Total† nutrients	Relative‡ salinity
sodium nitrate	16.5 N	100	16.5	100
ammonium nitrate	35 N	104.7	35.0	49.4
ammonium sulphate	21 N	69.0	21.0	53.7
ammonia solution	82 N	47.1	82.0	9.4
calcium nitrate	11.9 N, 17 Ca	52.5	28.8	30.1
urea	46 N	75.4	46	26.7
diammonium phosphate	21 N, 23 P	34.2	44	12.7
monoammonium phosphate	12 N, 27 P	29.9	39	12.7
superphosphate (single)	7.8 P	7.8	7.8	16.5
superphosphate (triple)	19.6 P	10.1	19.6	8.5
potassium chloride	49.8 K	116.3	49.8	38.5
potassium nitrate	13 N, 38 K	73.6	51	23.6
potassium sulphate	45 K	46.1	45	17.0
Kanit	14.5 K	109.4	14.5	124.5
calcium carbonate	40 Ca	4.7	40	1.9
calcium sulphate	23 Ca	8.1	23	5.8
magnesium oxide	60 Mg	1.7	60	0.5
magnesium sulphate (Kieserite)	16 Mg	44.0	16	44.5
Dolomite	24 Ca, 12 Mg	0.8	36	0.4

* The 'salt index' was calculated from the increase in osmotic *pressure* of *equal weights* of fertilizers.

† 'Total nutrients' have been recalculated from the sum of the N, P, K, Ca and Mg as usually stated in the fertilizer analysis, e.g. monoammonium phosphate = 12 N + 27 P = 39, but superphosphate = 7.8 P. Superphosphate is not usually regarded as being a source of plant calcium.

‡ 'Relative salinity' has been calculated from the increase in osmotic potential *per unit of plant nutrient*.

the fertilizer relative to its nitrogen phosphorus or potassium content. In both cases the results have been expressed relative to the effect of sodium nitrate. The actual increase in osmotic potential from a given amount of fertilizer will depend upon the medium, principally the amount of water and colloidal material it contains. It will be seen that on the *equal weight basis*, sodium nitrate and ammonium nitrate have approximately equal effects, with values of 100 and 104.7 respectively. When, however, comparison is made on an *equal nutrient basis*, which in this instance is nitrogen, then ammonium nitrate with a value of 49.4 has only half the salinity effect of sodium nitrate with a value of 100.

In addition to their direct effects on plant growth by way of the increase in the osmotic potential or salinity of the mix, some fertilizers may be less desirable than others because of an associated effect or the risk of a specific toxicity, e.g. chlorides in the case of potassium chloride, and free ammonia and nitrates in the case of urea.

6.10   PLANT MINERAL LEVELS

Horticulturists have long been accustomed to having their mixes and glasshouse border soils analysed at regular intervals, and the results used as the basis for subsequent nutrition and management. More recently, the value of knowing the levels of the macroelements within the plant, usually of the leaf tissue, has been recognized. The advantages of this technique are: .

(a) positive identification of mineral deficiencies and toxicities can be made;
(b) corrective treatment by way of liquid feeding or foliar sprays can be made often before the symptoms are fully developed and the plant is showing damage;
(c) nutrient levels in the tissue are more stable than those in the media – whereas the level of nutrients in the mix can show wide fluctuations because of the frequency and strength at which liquid fertilizers are applied, the level of macroelements in the leaves changes more slowly.

The limitations of this technique are:

(a) Normal or 'optimal' levels of minerals in the leaf tissue have not yet been established for the full range of ornamental crops.
(b) The amount of minerals present in the leaf is known to vary with the position of the leaf on the plant. Some elements, such as nitrogen, phosphorus, potassium and magnesium, are mobile, and when deficiencies occur in the mix, the plant is able to transfer these elements from the older leaves to the regions of active growth. Other elements, such as calcium, iron and boron, are immobile. Deficiencies of these elements are therefore first found in the regions of new growth.
(c) Differences in the 'optimal' mineral levels can vary also with the season, the age of the plant and the variety, as shown by the work of Nelson & Boodley (1966) with the carnation.

For these reasons, there must be careful standardization of the tissue chosen for analysis; usually the fifth and sixth leaves from the growing point are used.

The levels of macronutrients found to be satisfactory in the tissues of some ornamental plants are given in Table 6.8.

Table 6.8 Suggested levels of macroelements in the tissues of certain ornamental plants (% by weight of dry tissue). Values below which deficiency symptoms may start to appear are given in parentheses.

Crop	Nitrogen	Phosphorus	Potassium	Calcium	Magnesium
Azalea	2.0–3.0	0.2–0.5	1.0–1.6	0.45–1.6	0.2–0.5
	(1.8)	(0.15)	(0.75)	(0.2)	(0.2)
carnation	3.2–5.2	0.2–0.3	2.5–6.0	1.0–2.0	0.25–0.5
	(3.0)	(0.05)	(2.0)	(0.6)	(0.15)
Chrysanthemum	4.0–6.5	0.2–1.0	4.5–6.5	1.0–2.0	0.35–0.65
	(3.5)	(0.2)	(3.5)	(0.5)	(0.14)
geranium	3.3–4.8	0.4–0.67	2.5–4.5	0.81–1.2	0.2–0.52
	(2.4)	(0.24)	(0.7)	(0.77)	(0.14)
Poinsettia	4.0–6.0	0.3–0.7	1.5–3.5	0.7–2.0	0.4–1.0
	(3.0)	(0.2)	(1.0)	(0.5)	(0.2)
rose	3.0–5.0	0.2–0.3	1.8–3.0	1.0–1.5	0.25–0.35
	(3.0)	(0.2)	(1.8)	(1.0)	(0.25)
Aphelandra squarrosa	2.0–3.0	0.2–0.4	1.0–2.0	0.2–0.4	0.5–1.0
Ficus	1.3–1.6	0.1–0.2	0.6–1.0	0.3–0.5	0.2–0.4
Philodendron	2.0–3.0	0.15–0.25	3.0–4.5	0.5–1.5	0.3–0.6
Sansevieria	1.7–3.0	0.15–0.3	2.0–3.0	1.0–1.5	0.3–0.6

Based on work by Fortney & Wolf (1981) and Poole *et al*. (1976).

6.11   FOLIAR FEEDING

Macroelements are usually supplied as base or liquid fertilizers for plants to absorb through their roots. However, some elements can be assimilated from foliar sprays. This is often the best way of correcting magnesium deficiency, induced by very high levels of potassium in the medium, or a temporary calcium deficiency. A non-ionic wetting agent should be used with all sprays. The recommendations given in Table 6.9 are largely based on vegetable crops; because of the risk of scorching with some plants small tests should be made first.

Table 6.9 Macroelements that can be applied as foliar sprays.

Element	Salt	Rate	Remarks
magnesium	Epsom salts ($MgSO_4 \cdot 7H_2O$)	$20\,g\,l^{-1}$	more effective than soil applications
calcium	calcium nitrate ($Ca(NO_3)_2 \cdot 4H_2O$)	$2\,g\,l^{-1}$ $7.5\,g\,l^{-1}$	chrysanthemums tomato plants
nitrogen	urea ($CO(NH_2)_2$)	$2\,g\,l^{-1}$	tomato plants, soil application more effective
potassium	potassium sulphate (K_2SO_4)	$20\,g\,l^{-1}$	liable to scorch, soil application more effective

CHAPTER SEVEN

Microelements

Elements which are required by plants in only very small amounts are known as *micro-* or *trace* elements. Although they are present in plants at rates of only 10^{-3}–10^{-5} times the amount of such macroelements as nitrogen and potassium, they are nevertheless equally essential for normal plant growth. Deficiency of a microelement such as molybdenum can be just as important as macroelement deficiency. The elements which are generally regarded as being essential microelements are boron, chlorine, copper, iron, manganese, molybdenum and zinc.

Media based on mineral soils do not normally require any microelement additions, whereas those based on peat are often deficient in boron, copper, iron and molybdenum. The extent to which microelement deficiencies occur in peat-based mixes depends upon the species of plant being grown and also the type and source of peat, e.g. molybdenum deficiency occurs more frequently in lettuces, cauliflowers and poinsettias than it does in chrysanthemums. *Primula obonica* has a high requirement for copper whereas *Fuchsia* and *Rhipsalidopsis gaertneri* are susceptible to boron deficiency. Boron deficiency is also seen more frequently in plants grown in sphagnum than in sedge peats.

Plants are generally much less tolerant of high levels of microelements than they are of macroelements. For example, the amount of potassium in the mix can vary by several hundred milligrams per litre before plants show marked reactions, whereas with the microelement boron, deficiency can occur below 0.5 mg l^{-1} and toxicity at concentrations above 3 mg l^{-1}; copper, manganese and zinc can also be toxic at relatively low concentrations. In mixes made from materials having a cation exchange capacity, the cations copper, zinc and manganese behave in a similar way to other macronutrient cations such as calcium and potassium, and the anions boron and molybdenum behave similarly to phosphorus. With the exception of molybdenum, the availability of the microelements to plants decreases as the media pH is increased; the reverse effect is obtained with molybdenum. The effects which pH has on the availability of microelements in organic and lightweight potting mixes are contrasted with those of mineral soils in Figure 4.9. In lightweight mixes a pH range of 5.0–5.5 is generally preferred for maximum microelement availability.

Various solvents have been used to determine the availability of microelements in growing media. The DTPA method, which has been used for over a decade to estimate the availabilities of zinc, iron, manganese and copper in mineral soils (Lindsay & Norvell 1978), has also been used for peat and lightweight potting mixes (Markus *et al.* 1981). The DTPA extracting solution consists of:

0.005 M DTPA (diethylenetriaminepentacetic acid)
0.01 M CaCl₂ (calcium chloride)
0.1 M TEA (triethanolamine)

In a comparison of DTPA and EDTA as extractants for peat media, Haynes & Swift (1986) found with DTPA the availabilities of iron, copper and manganese decreased as the pH of the peat was increased. Paradoxically, with EDTA the availabilities appeared to increase with the pH of the peat; for this reason EDTA was not considered to be a suitable microelement extractant for peat. Broschat & Donselman (1985) used 1 M ammonium acetate to monitor the levels of extractable iron, manganese, zinc and copper in a peat, perlite and sand mix over an 18 month period. Hot water is generally used to estimate the amount of available boron; however, for molybdenum analysis of the leaf tissue is usually preferred to media analysis.

For most microelements determination of their concentration in the leaf tissue gives a clear indication of their status in the plant. With iron, however, the total concentration may not be a reliable indication of its activity or availability; often the amount of iron present in leaves showing deficiency symptoms is as high as in normal leaves. For this reason tests which measure the activity of specific enzymes in leaves have been used to assess the microelement status (Bar-Akiva 1964). Active iron status is measured by peroxidase activity, molybdenum by nitrate reductase and copper by ascorbic acid oxidase activity.

The microelement content of peat bogs has been studied in Scotland by Mitchell (1954) and in Ireland by Walsh & Barry (1958); typical values reported by Mitchell are given in Table 7.1. Walsh & Barry found that the

Table 7.1 Microelement content of some Scottish peats.

Peat source	Composition	Sample depth (cm)	% ash	Microelement (p.p.m. dry wt) Mo	Fe	Zn	Cu	Mn
Red Moss (Aberdeenshire)	younger sphagnum (*Calluna–Eriophorum*)	150–165	1.1	0.17	300	6	1.1	4.6
Aird's Moss (Ayrshire)	*Sphagnum–Eriophorum*	60	0.9	0.23	175	10	1.7	1.6
Westerdale (Caithness)	*Sphagnum–Scirpus*	15–45	2.3	0.60	580	10	6.3	2.5
Loch Chalium (Caithness)	*Sphagnum–Scirpus*	15–45	2.1	0.50	1120	8	13.2	4.2

microelement concentrations varied with the depth of the profile. The amount of elements in the top layers was usually greater than at a depth of 3 m, but large variations were found to exist both within and between bogs. They concluded that atmospheric precipitation of elements in sea-derived salts and in dust were important sources of minerals in bog development. Sedge peats in the UK contain greater amounts of microelements than sphagnum peats, this is largely due to the way in which they are formed, i.e. with drainage into the bog from the surrounding mineral soils.

The quantities of microelements naturally present in potting mixes are variable and will depend upon several factors. For example, bark from some species and locations can be high in manganese; some sands contain sufficient molybdenum for plants without any further additions; fly ash from fuel used for electricity generation can be high in boron. As well as the microelements naturally occurring in the nitrogen, phosphorus, potassium, calcium and magnesium fertilizers (§ 7.12), some are also present in the irrigation water, e.g. the high boron levels in some western parts of the USA.

7.1 BORON

Boron is associated with several functions in the plant, e.g. respiration, carbohydrate metabolism, the transport of sugars and the absorption of calcium. It is also essential for normal cell division and differentiation. Boron deficiency frequently causes the death of the apices, resulting in the production of numerous axillary shoots thereby giving a 'witches broom' effect. Deficiency symptoms vary somewhat with the plant species; in tomato plants, the leaves show a yellow chlorosis, sometimes with reddish-purple pigmentation and often accompanied by die-back from the tips; they are fleshy, brittle and have a high sugar content, and they may also show a red-brown colour in the veins when viewed from the underside. The fruit usually shows a ring of corky splits around the calyx. Celery shows numerous latitudinal splits or cracks on the underside of the stems. In chrysanthemums, the florets or petals are in-curved and quill-like; this is most noticeable in the broad petal reflexed flower varieties such as the Princess Anne group, as shown in Figure 7.1. In the carnation, the deficiency is frequently seen as the 'witches broom' effect. Antirrhinums show puckering of the leaves with yellow–gold chlorotic patches; the main shoot is often blind giving the plant a highly branched appearance.

Experiments have shown that plants grown in mixes having organic or ammonium-producing forms of nitrogen in the base fertilizer are more susceptible to boron deficiency than those having nitrate nitrogen (Bunt 1972). This is caused partly by the rise in pH which occurs as the organic nitrogen is first converted into ammonium before nitrification commences; also, the increase in micro-organism activity resulting from the use of organic fertilizers causes competition for the available boron. An

Figure 7.1 Boron deficiency in chrysanthemum flowers. Deficient flower on the left shows an in-curved form with the florets being quilled. Normal flower on the right.

example of the effect of the nitrogen source on the incidence of boron deficiency is seen in Figure 7.2. Antirrhinums grown with hoof and horn or urea as the nitrogen source showed typical boron deficiency symptoms, whereas plants grown with calcium nitrate were normal (see also Table 7.5). The uptake of boron can also be suppressed by high levels of phosphorus.

Figure 7.2 Plants grown with ammonium-producing fertilizers are more susceptible to boron deficiency. *Left*: antirrhinum in peat–sand mix with hoof and horn as the nitrogen source. *Centre*: urea as the nitrogen source. *Right*: calcium nitrate as the nitrogen source.

Symptoms of boron toxicity are first seen as a necrosis around the edges of the leaves. This is caused by boron being carried in the transpiration stream and then deposited at the extremity of the leaves where the localized concentration can be several times greater than in the central parts of the leaves; this symptom is shown in the chrysanthemum leaves in Figure 7.3. Plants such as *Beloperone guttata*, *Chrysanthemum*, garden pea, *Poinsettia* and *Zinnia* are very sensitive to boron toxicity, whereas the carnation and geranium are relatively tolerant.

Boron can be added to the mix with the base fertilizers, either as an inorganic salt, such as sodium borate or borax ($Na_2B_4O_7 \cdot 10H_2O$, having 11.3% boron), or in a general mixture of microelements in a fritted form. With crops that have a high boron requirement, such as carnations grown in glasshouse borders, boron can also be given regularly at 0.5 p.p.m. in the liquid feed. The limits between plant deficiency and toxicity levels with this element are very narrow. Bunt (1971) found that when chrysanthemums were grown in a peat–sand mix without any added boron, deficiency symptoms were produced in the flowers (see Fig. 7.1), and when 22 g m^{-3} (0.6 oz yd^{-3}) of borax were added toxicity symptoms developed in the leaves. A dressing of 7 g m^{-3} (0.25 oz yd^{-3}) of borax was found to be sufficient to correct the deficiency without causing toxicity symptoms. However, such low rates of application call for very thorough control of the mixing, and a much safer method of applying boron to mixes is to use the fritted form. This can be regarded as being a 'slow-release' form of boron with a low risk of toxicity; further discussion on the use of microelements in fritted form is given in section 7.9.

Figure 7.3 Boron toxicity in chrysanthemum leaves shows as a leaf edge necrosis.

Table 7.2 Weights of boron salts required to prepare stock solutions to give 0.5 p.p.m. of boron after dilution at 1 part in 200.

Stock solution, to be diluted 1 in 200.	Sodium borate $(Na_2B_4O_7 \cdot 10H_2O)$ (11.3% boron)	'Solubor' $(Na_2B_8O_{13} \cdot 4H_2O)$ (20.5% boron)
$g\,l^{-1}$	0.885	0.488
oz per 10 Imperial gal	1.4	0.78
oz per 10 USA gal	1.2	0.65

Boron can be safely used in liquid fertilizers only where an accurate and reliable means of diluting the stock solution is available, otherwise toxicity could soon occur. If boron is to be included in the liquid fertilizer, a form known as 'Solubor' can be used. This has a formula of $Na_2B_8O_{13} \cdot 4H_2O$ with 20.5% boron, i.e. it has almost double the boron content of borax, and its cold water solubility is about three times that of borax. *It is very important to remember which of these forms of boron is being used when calculating the amount required to make the stock solution.* The quantities of either of these two materials, required to give 0.5 p.p.m. of boron in the liquid feed, are given in Table 7.2. It is assumed that a stock solution is first made and then diluted at the rate of 1 in 200. For *regular* or *frequent* applications, the strength should not exceed 0.5 p.p.m. boron; where only an occasional application of boron is required to correct a deficiency, the strength can be increased to 2 p.p.m.

7.2 COPPER

Copper is involved in respiration and oxidation–reduction reactions in plants, functioning in several of the enzyme systems. Plants deficient in copper are low in ascorbic acid oxidase, a copper-containing enzyme. Copper is also necessary for the utilization of iron. In certain cases (Bunt 1971) chlorosis in chrysanthemum leaves has been controlled by the application of either copper or iron separately.

Deficiency symptoms in plants vary considerably. In fruit trees, it is seen as a rosetting of the terminal leaves; in lettuce plants there is failure to form hearts, with the leaves being narrow and cupped. Tomato plants show inward rolling of the leaves and stunted growth. With chrysan-themums, deficient plants have narrow leaves with interveinal chlorosis and the lower axillary shoots fail to develop; flowering is delayed by ten days in cases of mild deficiency, with a complete failure of the flowers to open in severe cases. Plants grown with high levels of nitrogen have a higher requirement for copper, especially when the nitrogen is in the ammonium form.

Copper deficiency frequently occurs in peat and organic soils, especially when the pH is below 4.5 and the level of extractable copper is below 3 mg l^{-1}. Vlamis & Raabe (1985) found plants of manzanita

(*Arctostaphylos densiflora*) were copper deficient even when 0.11 p.p.m. copper was given at each irrigation; it was concluded that the medium, consisting of two parts fir bark and one part sand, was fixing the copper. The deficiency can be corrected by applying 10–20 g m^{-3} of copper sulphate. Raising the pH value of very acid peats to 5.3 increases the availability of the copper, but at higher pH values availability decreases again. The natural chelates formed by copper in peat also increase its availability to plants.

Copper is another of the microelements which can cause phytotoxicity at high concentrations, toxicities having been reported in *Citrus* and *Gladiolus*. Often the roots of plants grown in media having very high levels of copper may contain up to ten times as much copper as the shoots. Excessive copper uptake usually reduces the iron content of the leaves and applications of iron chelate and phosphorus to the mix will reduce the effects of copper toxicity.

7.3   MANGANESE

Manganese is a constituent of certain enzyme systems concerned with respiration, nitrogen metabolism and the transference of phosphate. Although it is not a constituent of chlorophyll, it is found in high concentrations in the chlorophyll-containing tissue and a deficiency of manganese prevents chlorophyll formation. Manganese also affects the form in which iron is present in the leaves. When manganese is deficient, the iron is oxidized from the ferrous to the non-available ferric form and in some plants this is deposited in the leaf veins.

Deficiency symptoms may occur as a chlorotic marbling or as specks on either the young or old leaves, and is followed by necrotic spots, the small veins remaining green. Brassica crops are regarded as being particularly sensitive to manganese deficiency. Only the exchangeable and water-soluble forms of manganese are available to plants and, when new peat is limed and the pH raised to above 6.0, the amount of available manganese is very low and plants may soon show deficiency symptoms. The deficiency can be prevented by including with the base fertilizer 10–20 g m^{-3} of manganese sulphate.

High levels of available manganese in the mix can be toxic to young seedlings. Symptoms of manganese toxicity in tomato seedlings are first seen in the cotyledons, which turn yellow and die early. Older leaves point downwards and the leaflets curl under; the veins show a dark purplish-brown necrosis with some brown spotting occurring in the adjacent tissue. The young leaves show chlorosis and the seedlings have a generally stunted appearance. Manganese toxicity can occur under four general conditions:

(a) anaerobic conditions in the medium caused by poor drainage;
(b) following steam pasteurization of certain types of soil;
(c) under acid conditions;
(d) the use of materials naturally high in manganese, e.g. some barks.

Needham (personal communication) has reported toxicity symptoms in the Princess Anne varieties of chrysanthemum when grown in a peat–sand mix having a pH of 4.9 with an extractable manganese level of 24 mg l^{-1}. Leaves of the affected plants contained 1600 p.p.m. of manganese; the peat had not been steam sterilised. Raising the pH of the medium with lime removes the toxicity by reducing the amount of available manganese.

7.4   MOLYBDENUM

Molybdenum is required by plants for the reduction of nitrate nitrogen to ammonia within the plant before proteins can be formed. The quantity of this element necessary for the normal functioning of the leaves is very low, even by microelement standards; often the molybdenum level in healthy leaves is less than 1 p.p.m. of the dry weight.

Molybdenum is required in such small quantities that plants which are vegetatively propagated, such as chrysanthemums, frequently fail to show deficiency symptoms if the cuttings have been taken from stock plants which were adequately supplied with the element. Certain seed-propagated plants such as lettuce and brassicas are very susceptible to the deficiency. The most frequently observed symptoms of molybdenum deficiency are pale coloured leaves followed by marginal scorching. Tomato leaflets show loss of chlorophyll with an upward curling, followed by withering commencing from the apical leaflet. The outer leaves of lettuce seedlings turn pale and then die. Probably the best known symptom of molybdenum deficiency is 'whiptail' of cauliflower, which occurs as a severe restriction on the width of the leaf blade; in some cases the leaf may consist of little more than the midrib.

Results of an experiment made on the molybdenum requirement of lettuce seedlings (Bunt 1973a) show that a response was obtained from the addition of 5 g m^{-3} of ammonium molybdate to the seed sowing mix (Table 7.3). A further response in plant growth was obtained when the

Table 7.3   Response of lettuce, cv. Kwiek, to molybdenum added to the seed sowing and potting mixes (fresh weight in g).

Seed sown in peat–sand seed sowing mix	Seedlings grown in peat–sand potting mix with frit 253A at 375 g m^{-3}		Seedlings grown in John Innes potting compost (JIP-1)
	no extra molybdenum	Extra molybdenum, 5 g m^{-3} of ammonium molybdate	
no molybdenum	11.2	32.6	26.2
with 5 g m^{-3} of ammonium molybdate	23.2	33.1	27.6

Frit 253A contains: 2% boron, 2% copper, 12% iron, 5% manganese, 0.13% molybdenum, 4% zinc.

ammonium molybdate was added to the potting mix, even though the potting mix contained a fritted trace element mixture which included some molybdenum. The quantity of molybdenum supplied by this fritted trace element mixture was obviously not sufficient, and subsequent to this work a new formulation of frit with a higher molybdenum content has been made (§ 7.9).

Primulas also appear to be sensitive to molybdenum deficiency; plants grown in winter have been reported as having a greater molybdenum requirement than summer-grown plants (Reeker 1960). Those plants receiving nitrogen in the nitrate form are more likely to show deficiency symptoms than plants receiving the ammonium form. High levels of sulphates in the mix can also reduce the uptake of molybdenum by plants.

The deficiency can be controlled by including with the base fertilizer either ammonium or sodium molybdate at a rate of 2.5 g m^{-3} of mix; toxicity has been reported at a rate of 100 g m^{-3}. Molybdenum can also be used in the fritted form; the mixture of fritted trace elements should contain not less than 1% by weight of molybdenum. Poinsettias can be very susceptible to this deficiency when propagated and grown under certain conditions, and, as a safeguard, molybdenum is often included at 0.1 p.p.m. in the liquid feed. This strength can be obtained by dissolving 5 g of sodium molybdate (Na$_2$MoO$_4 \cdot$2H$_2$O, 39.7% molybdenum) in 100 l of stock solution and diluting at 1 in 200 (4.6 g per 20 Imperial gal or 3.8 g per 20 USA gal, and dilute at 1 in 200).

7.5 IRON

Varying degrees of iron deficiency are often seen in certain ornamental plants growing in peat-based substrates. Amongst the ornamentals most prone to this problem are: *Azalea, Bougainvillea, Calluna, Camellia, Erica, Gerbera, Hydrangea, Magnolia, Meconopsis, Petunia, Primula, Rhododendron* and rose. The vegetable crops commonly raised in potting mixes are less prone to iron deficiency; however, it can sometimes be seen in tomato plants when they are fruiting. Despite intensive research, the function of iron in plants is not clearly understood; it is not a constituent of chlorophyll but it is essential for its formation. Iron is also a constituent of certain enzymes, e.g. peroxidase and catalase.

It is present in soils as Fe^{3+} (ferric ion) and has to be reduced to Fe^{2+} (ferrous ion) by the plant roots before it is absorbed. Sometimes the deficiency is caused by the inability of plant roots to absorb the iron, in other cases the absorbed iron is precipitated or inactivated after entering the plant. In a comprehensive review, Wallace & Lunt (1960) listed several causes of the deficiency, which can be grouped conveniently under two headings.

Physical conditions in the substrate, these include:

(a) over-irrigation or wet conditions;
(b) low aeration;
(c) low soil temperature.

Chemical balance within the substrate, this includes:

(a) low iron supply;
(b) high calcium and bicarbonates resulting in high pH;
(c) high phosphorus;
(d) high levels of zinc, manganese and copper;
(e) high nitrate:ammonium ratio.

It is also known that some plants are more subject to the disorder when grown under high light conditions, e.g. *Nandidia compacta*, or at low temperatures. Container-grown plants are particularly liable to root damage by over-watering and by over-fertilization, especially if strong liquid feeds are given when the medium is dry. This damages the root hairs and restricts their ability to reduce Fe^{3+} to Fe^{2+}. This ability is also impaired by high levels of phosphorus, copper and nickel. An example of root damage resulting in iron deficiency is shown in Figure 7.4. High ammonium concentration following the use of an organic source of nitrogen caused root damage, this resulted in nine leaves being produced with acute iron deficiency symptoms. After a period of recovery, during

Figure 7.4 Chrysanthemum plants showing iron deficiency caused by root damage from high ammonium levels. Note the normal green leaves produced before and after the period of root damage.

which new roots were formed, normal growth was resumed without any corrective measures being taken. Plants that are kept wet when the rate of evapotranspiration is low are also liable to show deficiency symptoms.

Some plants are more prone to show iron deficiency when they are given only nitrate nitrogen. One explanation of this is the secretion by the plant of bicarbonate, which raises the pH around the roots and thereby reduces the availability of iron. Above pH 4 the activity of iron in solution decreases a thousandfold for each unit rise in pH. The reduced activity of iron within the plant as result of nitrate absorption is also thought to be partly due to its binding with the larger amount of organic acids formed with nitrate nutrition (§ 5.2). Organic-based potting media, which do not show anion fixation as do mineral soils, often have high levels of soluble phosphorus present, this can result in iron phosphate being precipitated on the roots and, therefore, unavailable to the plant. There is also evidence that high levels of phosphorus within the plant can cause a deficiency by inactivating the iron; the total amount of iron in deficient plants is often as high as in normal plants and the ratio of phosphorus to total iron may show a negative correlation with the amount of *active* iron in the leaves.

Sometimes bark that has not been composted will have high levels of available manganese (§ 2.2.3), this can cause iron deficiency as well as manganese toxicity. Read & Sheldrake (1966) reported iron deficiency in petunias grown in peat–vermiculite mixes, it was attributed to the high Mn:Fe ratio in the vermiculite. Leaves of plants showing iron deficiency are sometimes higher in potassium and lower in calcium than normal leaves; however, this may be the result of the chlorosis rather than the cause.

Although the soluble iron content of the acidic peats is normally greater than that of neutral or alkaline mineral soils, iron deficiency can occur in plants grown in the younger sphagnum peats. This can be corrected by applying ferrous sulphate or an iron chelate at 25 g m^{-3} of mix ($\frac{3}{4}$ oz yd)$^{-3}$ with the base fertilizer; the chelated form of iron will usually give a better result. When symptoms first appear in the plants, a 0.1% solution of iron chelate applied to the mix will usually be more effective than a foliar spray.

Chemical diagnosis of iron deficiency by measuring the peroxidase activity in leaves will often be a more sensitive and reliable method than determining the total iron present. For example, Besford & Deen (1977) found that the peroxidase activity in leaves of conifers receiving iron in the liquid feed was increased 11.5 fold in comparison with the control plants, whereas the increase in the total iron was only 2.6 fold. Chelated iron added with the base fertilizer resulted in a 29 fold increase in enzyme activity, whereas the total iron content increased only 7.5 fold. The use of chelated microelements is discussed in Section 7.10.

7.6 ZINC

Zinc deficiency is not generally encountered in pot-grown plants. It has, however, been reported in Florida in plants grown on organic soils. The deficiency has occurred in citrus and peach trees and is known as 'mottle

161

leaf' and 'little leaf'. A small quantity of zinc sulphate in the range 0.5–5.0 $g m^{-3}$ is often included with the base fertilizer as a precautionary measure.

Zinc toxicity in plants can sometimes occur when new galvanized (zinc-coated) pipes are used in the water supply system. Tests have shown that the first few gallons of water drawn off in the morning, after water has been static in the pipes overnight, can contain as much as 4 p.p.m. of zinc. This value soon falls as the pipes are flushed with fresh water and the zinc level will then fall to 0.5 p.p.m. or less. When the pipes have aged, the zinc value will often be not greater than 0.1 p.p.m. Static water drawn from untreated galvanized tanks can contain as much as 10–20 p.p.m. of zinc. In water culture studies it has been found that the zinc level should not exceed 0.2 p.p.m. Localized zinc toxicity of plants can also occur from condensation dripping from newly galvanized wire and tubular blackout supports such as are used in chrysanthemum growing.

7.7   CHLORIDE

The requirement of plants for chloride is very low and it can be shown to be an essential element only by growing plants in water cultures and working under closely controlled conditions. Chlorides usually occur in sufficient quantity in the irrigation water and as impurities in the fertilizers to supply all the plant's requirements. Normally, chlorides are of concern only when they reach toxicity levels. This shows as a burning or scorching of the leaves, together with bronzing and premature defoliation. In some parts of the world, chlorides are present in the irrigation water at sufficiently high concentrations to cause toxicity, i.e. >70 p.p.m. Cl^-. In England, the chloride content of the public water supply usually does not exceed 20 p.p.m. Similarly with free chlorine, public water supplies do not normally contain more than 0.5 p.p.m. of free chlorine. Water containing 2 p.p.m. of free chlorine is not drinkable and plants have been given water with 5 p.p.m. of free chlorine without showing any toxic symptoms.

American workers have examined the reaction of a number of ornamental plants to chlorides and found *Gardenia*, geranium and *Poinsettia* to be particularly sensitive. Stocks were classified as being intermediate and carnation, *Penstemon* and *Verbena* were regarded as being tolerant to chlorides. Arnold Bik (1970), in the Netherlands, examined the effect of chloride and nitrogen levels on the growth and chemical composition of *Gloxinia* and *Chrysanthemum*. He found that the detrimental effect of chloride was primarily due to the related increase in the salinity as measured by the electrical conductivity of a saturated medium extract, i.e. the EC_e; there was also some evidence of a specific ion effect. *Gloxinia*, which is a salt sensitive plant, showed a decline in growth when the EC_e rose above 4 mmho cm^{-1}, while for the chrysanthemum, which is a more tolerant plant, the corresponding value was 6–8 mmho cm^{-1}.

Table 7.4 Effect of soil pH on the colour and aluminium content of hydrangea flowers.

pH range	Colour	Aluminium (p.p.m.)
4.6–5.1	blue	2375–897
5.5–6.4	blue, tinged pink	338–187
6.5–6.7	pink, tinged blue	214–201
6.9–7.4	clear pink	<217

7.8   ALUMINIUM

This element is not essential to plant life and normally receives attention only when a toxicity arises, e.g. in plants grown in very acid soils. Many plants accumulate aluminium, e.g. the tea plant.

The hydrangea is of interest because the colour of its flowers is dependent upon the amount of aluminium they contain. In its native habitat this plant produces flowers (sepals) which are naturally blue. In some locations, however, the flowers are often pink. This phenomenon was investigated by Allen (1943), who grew plants in soils with pH values ranging from 4.6 to 7.4, and related the pH of the soil to the colour and aluminium contents of the flowers; his results are summarized in Table 7.4. By the use of foliar sprays and the absorption of salts through cut stems, he showed that the flower colour was dependent upon the amount of aluminium in the tissue. Iron salts had no direct effect upon the colour but they had an indirect effect by reducing the pH, which in turn increased the availability of the aluminium. However, iron salts are beneficial in preventing iron chlorosis, a deficiency to which this plant is particularly susceptible.

To produce blue hydrangea flowers it is customery to grow the plants in a medium with a pH of 4.8–5.0; above this range there is a marked reduction in the amount of aluminium in solution. Aluminium sulphate used at 3 kg m^{-3} (5 lb yd^{-3}) in addition to the normal fertilizers helps to maintain a low pH and also provides sufficient soluble aluminium. When the plants are brought into the glasshouse in January for forcing, after having had a period of low temperatures to break their bud dormancy, a 1% solution of aluminium sulphate is given at 14-day intervals. High levels of phosphorus in the mix will result in the formation of insoluble aluminium phosphates and for this reason plants which are to be 'blued' are given only low levels of phosphorus. When pink flowers are required, the soluble aluminium must be kept at a low level by growing at pH 6.5, omitting all aluminium sulphate treatments and maintaining high levels of phosphorus.

7.9   FRITTED MICROELEMENTS

One problem encountered in supplying microelements to mixes is the very narrow limits which exist between deficiency and toxicity levels; as previously shown, boron toxicity occurred when sodium borate was used

163

at rates above 7 g m^{-3}. Applying such small quantities of a chemical presents problems in the weighing and uniform mixing into the media. Boron is also quite soluble in organic media such as peat and when plants require frequent watering there is some risk of its loss by leaching, with a deficiency occurring later on. One method of overcoming this problem is to supply boron in a slowly available form such as a soft glass frit; in this way the safety margin is increased and the risk of leaching decreased. Glass frits are used in industry in the manufacture of porcelain enamel and ceramic glaze finishes. For agricultural purposes, frits, which are known as FTE, i.e. fritted trace elements, are prepared by adding the desired amounts of inorganic salts to sodium silicate. This is heated to a temperature of around 1000 °C and the molten mass is poured on to either cold steel or running water, where it cools and shatters into small fragments which are ground into a very fine powder. The rate at which the frits dissolve in the media and release their microelements to plants can be controlled by adjusting the rate of cooling, the fineness of the grinding and the addition of chemical impurities. Some of the properties of frits have been reviewed by Holden *et al.* (1962). Frits are considered a satisfactory method of supplying boron to plants but they are less efficient in supplying iron.

Once the fritted trace element mixture has been prepared, the rate at which minerals are released is primarily controlled by the pH of the mix. Bunt (1972) examined the effects of the pH of the mix and the rate of application of a fritted trace element mixture (FTE 253A) containing 2% B, 2% Cu, 12% Fe, 5% Mn, 0.13% Mo, 4% Zn, on the growth and chemical composition of cauliflower, celery, chrysanthemum, tomato and pea plants. Large interactions were found among the plant species, the pH of the mix and the rate of frit. Tomato plants grown in mixes to which no frit had been added showed a marked response to the pH, the optimal pH being 6.4 (in 0.01 m CaCl$_2$) with marked depressions in growth occurring above and below this value (Fig. 7.5). When the frit was added at the rate of 350 kg m^{-3} (10 oz yd^{-3}), the effect of the mix pH was largely eliminated and also plant growth increased significantly; there was very little difference in plant growth between the 350 kg m^{-3} and 1.75 kg m^{-3} rate (3 lb yd^{-3}). With chrysanthemums, however, the 1.75 kg m^{-3} rate suppressed growth. This was caused by the amount of boron released from the frit, chrysanthemum plants being much more sensitive to boron toxicity than the tomato plants.

As a result of this work, and also the experiment on the molybdenum requirement of lettuce (§7.4), Bunt (1947b) formulated a new frit having the following composition:

1% B, 4.3% Cu, 13.8% Fe, 5.4% Mn, 1% Mo, 4.3% Zn.

This frit has a lower boron content and also greater amounts of copper and molybdenum than the 253A frit. It has been used successfully at 350 g m^{-3} (10 oz yd^{-3}) for raising bedding plants, chrysanthemums, poinsettias, lettuce, tomato and other pot plants. Where there is likely to be a very heavy demand for boron, e.g. if tomato plants are cropped with several trusses of fruit in a 25 cm (10 in) pot, it may be necessary to give an occasional boron feed as the plants develop.

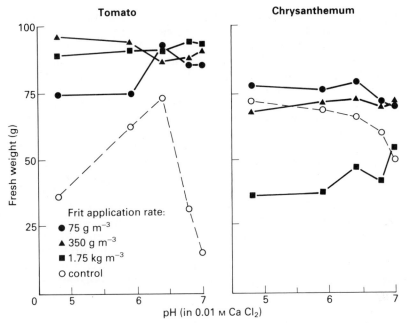

Figure 7.5 The effect of varying rates of a fritted trace element mixture at different pH values on the growth of tomato and chrysanthemum plants.

The source of nitrogen can also affect the rates at which microelements in the frit are absorbed by the plant (Bunt 1972). When chrysanthemums were grown with hoof and horn in the base fertilizer and urea as the nitrogen source in the liquid feed, the levels of boron and manganese in the leaf tissue declined sharply as the pH of the medium was raised by liming. In contrast, plants grown with ammonium nitrate as the nitrogen source in the base fertilizer and liquid feed showed much less effect of medium pH on boron and manganese levels (Table 7.5). The form of

Table 7.5 Effect of nitrogen source and pH on microelements in the leaves of chrysanthemums grown with frit 253A at a rate of $1.75 \, g \, l^{-1}$.

Nitrogen source	Calcium carbonate $(g \, l^{-1})$	pH (in $0.01 \, M$ $CaCl_2$)	Microelement (p.p.m. in dry leaf tissue)		
			boron	manganese	copper
hoof and	2	4.8	237	900	34
horn	4	5.9	190	644	26
	6	6.4	142	600	31
	8	6.8	73	444	26
	10	7.0	53	413	23
ammonium	2	4.8	280	869	36
nitrate	4	5.9	296	775	40
	6	6.4	280	750	31
	8	6.8	306	731	35
	10	7.0	253	600	32

nitrogen has a large effect on the media pH. In peat-based mixes the ammonification of organic fertilizers proceeds more rapidly than nitrification (§ 5.3.1), resulting in a temporary increase in pH. Adding more lime not only raises the pH of the mix, it also increases the rate of ammonification of the hoof and horn, this raises the pH further and thereby reduces the availability of the boron and manganese. Availability of the copper was apparently unaffected by the increase in pH. Similar effects on microelement availability have been observed on plants grown with base fertilizers and no liquid feeding.

The solubility of boron in the frit can also be affected by the water content of the media. Tomato plants propagated early in the year are sometimes grown 'dry' to avoid producing soft, vegetative plants that may abort flowers in the first inflorescence under the poor light intensities. Plants grown in this way may show mild boron deficiency symptoms in the lower leaves whereas plants grown 'wet' do not show the symptoms. Roll-Hansen (1975) found that although tomato plants grown with a frit containing 1.8% boron, used at rates up to 800 g m^{-3}, responded favourably with no symptoms of toxicity, cucumbers grown with 200 g m^{-3} of the frit showed toxicity symptoms. As a result a new frit, no. 36, was formulated:

$$0.5\% \text{ B, } 2.0\% \text{ Cu, } 9.0\% \text{ Fe, } 2.0\% \text{ Mn, } 0.5\% \text{ Mo, } 2.0\% \text{ Zn}$$

and is used at 200 g m^{-3}. The rate is increased to 500 g m^{-3} for long term crops and if the volume of growing medium per plant is small. In the UK, Frit 253A, used at 0.35 kg m^{-3}, has provided sufficient microelements for a two-year crop of carnations grown in peat.

Smilde (1971) grew tomato, lettuce and chrysanthemum plants in a 'black' peat, i.e. a humified old sphagnum peat formed from *Eriophorum*, *Carex* and *Phragmites*. Frits supplying boron, molybdenum and copper were efficient as single salts; there was no response to other minor elements in either form. Although there are some reported instances of iron in frit form being beneficial, it has often failed to correct the deficiency and it must be concluded that chelated iron is preferable.

7.10   CHELATED MICROELEMENTS

Often microelement deficiencies in plants are the result of an unavailability of an element or antagonisms between elements, rather than a true deficiency of the element. Iron is an example of an element rendered unavailable to plants because of its chemical precipitation in the medium by other elements. To overcome this problem, certain plant nutrients can be treated in a way which keeps them in an available form; the term used to describe this process is 'chelation' and is derived from the Greek word meaning 'claw'.

To be successful, a chelating material must:

(a) prevent the iron from being precipitated in the soil, e.g. by the action of high pH or phosphorus levels;
(b) prevent its replacement by other metals, e.g. copper;
(c) be non-phytotoxic;
(d) resist decomposition by soil organisms.

Examples of naturally occurring chelate materials are citrate and tartrate; recently synthetic chelates have been successfully prepared from the polyamino-polycarboxylic acids. One of the earlier materials used with success was EDTA (ethylenediamine tetraacetic acid) and several other chelating agents have been developed since.
Some of the more common synthetic chelating materials are:

Abbreviation	Chemical name
EDTA	ethylenediamine tetraacetic acid
DTPA	diethylene triamine pentaacetic acid
HEEDTA	hydroxyethyl ethylene diaminetriacetic acid
EDDHA	ethylenediamine-di(o-hydroxyphenylacetic) acid

When used for chelating iron, the soluble iron contents range from about 14% for EDTA to 6% for EDDHA.
Experience has shown that the benefits from using chelated iron depend upon the plant species, the type of medium and its pH. Most of the literature on the use of chelates refers to mineral soils, and although chelates are also effective in peats and lightweight mixes there are some differences in their behaviour in the two types of media. Boxma (1981) examined the stability of four iron chelates, Fe-EDTA, Fe-DTPA, Fe-HEEDTA and Fe-EDDHA, in sphagnum peat over the pH range 4.35–7.85. At pH values of 5.65 and below, more than 90% of the iron applied as Fe-EDTA and Fe-DTPA had remained water-soluble after 42 days; this is much greater than in mineral soils. At pH 7.25 and above, the iron in Fe-EDTA was largely insoluble after 3 days, and less than 30% of the iron in Fe-DTPA was soluble after 12 days. With Fe-HEEDTA only 50% of the iron was water-soluble after 42 days at pH 4.35, and this value decreased sharply as the pH was increased. A significant amount of Fe-EDDHA was sorbed on peat at low pH, only 65% was water-soluble after 42 days at pH 4.35; this contrasts with mineral soils where it is not fixed at low pH. As the pH of the peat was increased the percentage of soluble iron in Fe-EDDHA increased, being 86% at pH 5.65 and 90% at pH 7.85. From this and other studies, the suitabilities of the four chelates for use in peat-based potting mixes can be summarized:

Fe-HEEDTA: the water solubility is low and declines as pH is increased – not recommended.

Fe-DTPA: has good water solubility below pH 6, but loses its stability when mixed with fertilizers containing other microelements, particularly copper – therefore not recommended.

Fe-EDTA: high availability below pH 6, stable with other microelements – recommended for media below pH 6.0.

Fe-EDDHA: highest availability at pH 5.5 and above, stable with other microelements – can be used for most situations with peat and lightweight mixes.

However, some sources of Fe-EDDHA are less pure and therefore less effective than others. Luit & Boxma (1981) found that only one third of the iron in one commercial source was present as Fe-EDDHA. This source was also much less effective than pure Fe-EDDHA in controlling chlorosis in *Chaenomeles superba*.

Chelates can be included with the base fertilizer, usually at the rate of 25 g m^{-3}, or applied as a 0.1% solution to the potting medium when the plants are growing. They should not be used with liquid feeds that contain either ammonia or phosphoric and nitric acids. High pH causes precipitation and at low pH H^+ ions compete with the iron for the donor groups. Chelated iron can also be given as a foliar spray at a rate of 0.5 g l^{-1}; there is less risk of foliar damage if the Fe-EDTA form is used but some plants, e.g. primrose, are very susceptible to scorching.

Although iron is the most widely used of the chelated metals, other microelements such as zinc, copper and manganese can be chelated also, and general mixtures of microelements in EDTA form are available. Neither boron nor molybdenum can be chelated, however, and they can be included in the mixture only as inorganic salts, usually as sodium borate and ammonium or sodium molybdate; they can, of course, be applied separately in fritted form. It is important to remember that frits are a way of supplying microelements in a slowly available form, whereas chelated microelements have an immediate availability.

Experiments at the Glasshouse Crops Research Institute with a microelement mixture based on an EDTA chelate and having an analysis of:

$$1.7\% \text{ Cu}, 3.35\% \text{ Fe}, 1.7\% \text{ Mn}, 0.6\% \text{ Zn}, 0.875\% \text{ B}, 0.023\% \text{ Mo},$$

showed the optimal rate of application to be low; it also varied with the plant species. For *Zinnia*, the optimal rate was 75 g m^{-3} (2 oz yd^{-3}) whereas for aubergine and *Exacum* it was 150 g m^{-3} (4 oz yd^{-3}). At double this rate, there was a suppression of growth, the fresh weights being approximately one sixth that of the plants having the optimal rate.

7.11   INORGANIC SALTS AND PROPRIETARY FERTILIZERS

The microelement additions necessary to grow plants in soil-less media can also be applied as inorganic salts. Because they are soluble and immediately available, care must be taken to avoid toxicities from either

Table 7.6 Weight of inorganic salts to supply microelements to peat-based mixes.

Inorganic salt	Rate $(g\,m^{-3})$	Amount of microelement in mix $(mg\,l^{-1})$					
		boron	copper	manganese	zinc	iron	molybdenum
sodium borate	8	0.9					
copper sulphate	25		6.4				
manganese sulphate	25			5.0			
zinc sulphate	25				5.7		
iron sulphate*	50					10.0	
sodium molybdate	2.5						1.0

* This can be replaced by 25 g m^{-3} of chelated iron.

high rates of application or uneven mixing. Several recipes have been published and differences between them can be attributed in part to variations in the natural levels of microelements present in local materials, fertilizers or water supplies, and in part to the requirements and tolerances of local crops. The rates given in Table 7.6 are suitable for young peats with low microelement contents.

To avoid difficulties in weighing and mixing such small amounts of chemicals one of the proprietary microelement mixtures can be used (Table 7.7), alternatively some of the compound NPK fertilizers formulated for preparing loamless mixes have microelements included.

Table 7.7 Examples of proprietary microelement fertilizers used in soil-less potting mixes.

Fertilizer	Manufacturers' recommended rate	
CHEMEC	62.2 g m^{-3}	chelate, no molybdenum
Esigram	2.2 kg m^{-3}	elements sorbed to calcine clay granules, slow release
Micromax	1.0 kg m^{-3}	mostly sulphates
Perk	2.2 kg m^{-3}	oxides and sulphates
frits 503, 555	62.2 g m^{-3}	slow release, available USA
frits 253A, 255	375 g m^{-3}	slow release, available UK
Librel BMX	28–58 g m^{-3}	chelated, except boron and molybdenum
STEM		soluble, added to liquid feed at 8 oz 100 lb^{-1} fertilizer (0.5%)

7.12   OTHER SOURCES

Apart from the sources of microelements already discussed, indirect sources of supply are impurities in the water, e.g. zinc and fungicide sprays, e.g. copper. The most important single source of supply, however, is the impurities present in the commercial grades of fertilizers, of which the phosphorus fertilizers are the most heavily contaminated. Some indication of the amounts of microelements present in commercial fertilizers (Swaine 1962) is given in Table 7.8. The microelement values given in the table are for a typical range of concentrations found in the fertilizers; because of the very large variation possible between samples no attempt has been made to give either the highest or mean values. It is apparent from the data that it would be impossible to estimate the

Table 7.8   The microelement content of fertilizers.

Fertilizer	Microelement content (p.p.m.)				
	boron	copper	manganese	molybdenum	zinc
ammonium sulphate	0.2–25	0–20	tr.–80	tr.–0.2	0–100
ammonium nitrate	0.4–2.0	tr.–1	<5	0.1–0.3	1–5
urea	0–10	0–4	1–10		0–50
calcium nitrate	tr.–90	1–20	1–10		<1.0
sodium nitrate	50–300	1–25	<1	0.1	1–10
superphosphate (single)	3–15	10–60	10–200	tr.–10	70–500
superphosphate (triple)	tr.–200	30–200	0–200	3–20	0–100
monoammonium phosphate	10–100	10–100	30–200	2–10	30–200
phosphoric acid	<6	15–100	40–2000	100	1–300
phosphate rock	<50	1–30	10–200	tr.–20	5–300
basic slag	20–1000	10–60	1000–50 000	tr.–10	3–30
potassium nitrate	1–2	tr.–30	tr.–8		<8
potassium sulphate	<30	1–10	tr.–50	0.1–0.3	0–6
potassium chloride	0–150	0–10	tr.–8	tr.–0.2	<3
limestone	2–50	2–200	10–700	2–20	3–200
calcium carbonate	<0.3	0–50			3–30
Dolomite limestone	1–25	1–100	10–500	1–20	5–100

These data indicate only the range of values that are commonly present in commercial fertilizers. The actual amounts of trace elements present depend upon the method of fertilizer manufacture, e.g. whether it is made by a synthetic process or by treatment of raw fertilizer materials, such as rock phosphates; in the latter case the trace element concentration also depends upon the source of the phosphate. Rock phosphates obtained from American sources appear to contain more microelements than those from North Africa.

Table 7.9 Average microelement content of phosphorus fertilizers in the USA (from Bingham 1959).

Fertilizer	Microelement content (p.p.m.)				
	boron	copper	manganese	molybdenum	zinc
phosphoric acid	25	62	331	20	989
superphosphate (single)	57	70	48	18	1000
superphosphate (triple)	60	70	160	20	1500
ammonium phosphate	51	56	149	14	1054

quantity of microelements being supplied by the fertilizers unless an analysis of the actual materials was made. Some average values of the microelement content of phosphorus fertilizers are given in Table 7.9.

7.13   MICROELEMENT AVAILABILITY

Apart from those occasions when there is a true deficiency of microelements in the mix, deficiency symptoms in plants can be the result of:

(a) incorrect pH,
(b) imbalance or antagonism with other microelements,
(c) antagonism with macroelements,
(d) media too wet or too dry.

Several antagonisms between elements are known to occur and the most important factors controlling the availability of microelements are given in Table 7.10. It will be seen that phosphorus can have a significant effect upon the availability of several microelements. Bingham & Garber (1960) compared the effects of several types of phosphorus fertilizers upon microelement availability and obtained essentially the same results with all of the forms of phosphorus tested.

7.13.1   Microelements in plant tissue

Advisory Officers often find that a microelement analysis of the leaves of plants assists in cases where a microelement deficiency is suspected; the analysis also provides information upon which to base nutritional programmes. However, plant species show considerable differences in the concentrations of microelements present in their leaves and also in the levels at which deficiencies and toxicities become apparent. For example, boron deficiency in leaf tissue has been reported at the following levels: geraniums 18 p.p.m., roses 30 p.p.m. (Fortney & Wolf, 1981); *Tagetes* and *Coleus* 5 p.p.m., *Petunia* and *Salvia* 11 p.p.m., *Impatiens* 44 p.p.m. and *Cosmos* 62 p.p.m. (Johnson 1973). Similarly, the level at which boron toxicity occurs has been shown to vary with the crop: carnation and geranium >700 p.p.m., *Poinsettia* >200 p.p.m., *Chrysanthemum* >125 p.p.m.

Table 7.10 Conditions affecting availability of microelements in mixes.

Element	Conditions affecting its availability
boron	Availability reduced at pH >5.5 in peat, relatively easily leached from organic matter, e.g. peat. Organic nitrogenous fertilizers reduce availability, by temporary biological lockup and by rise in media pH during first stages of mineralization. High phosphorus levels reduce uptake by plants grown in acid media.
manganese	Availability much increased at low pH. Plant uptake reduced by high potassium levels and also high levels of iron, copper and zinc. Phosphorus application helpful in reducing plant uptake in cases of toxicity following steam sterilization, providing that pH is not thereby reduced. Iron chelate also useful in cases of toxicity. Ammonium ions have favourable effect in alleviating manganase toxicity in beans and flax.
copper	High rates of nitrogen, especially in ammonium form, increases copper deficiency, also high levels of phosphorus. Copper toxicity reduced by iron chelate and molybdenum.
iron	Deficiency induced by high levels of copper, manganese and zinc. High levels of phosphorus result in insoluble iron phosphates. 'Lime induced' chlorosis caused by over-liming and irrigation water high in carbonates and bicarbonates.
molybdenum	Availability decreased at pH <6.0, also by high levels of available manganese, and by SO_4^{2-} ions and copper. Plants receiving nitrate form of nitrogen have greater requirement than those having ammonium nitrogen.
zinc	Availability reduced at high pH and by applying magnesium carbonate; also high levels of phosphorus reduce availability and plant uptake.

(Fortney & Wolf 1981), cucumber >270 p.p.m., tomato >160 p.p.m. and lettuce >55 p.p.m. (Roorda van Eysinga & Smilde 1981). The microelement values given in Table 7.11 must, therefore, be regarded only as guide lines. As with the macroelements, the concentration of microelements often varies with the position of the leaf on the plant, and a standard procedure in collecting the tissue must be followed.

Table 7.11 Average microelement ranges found in the leaf tissue of ornamental plants.

Plant condition	Microelement range (p.p.m. in dry matter)				
	boron	manganese	copper	molybdenum	zinc
deficient	<20	<20	<5	<0.1	<15
normal	30–80	30–200	10–25	0.1–3	30–50
excessive	>150*	>800†	>70	>70‡	>200

* Carnations and roses are more tolerant of boron, toxicity may not occur below 500 p.p.m.

† Manganese toxicity occurs in glasshouse-grown lettuce at 500 p.p.m. manganese.

‡ There is very little data available on molybdenum toxicity, 20 p.p.m. is regarded as excessive in cucumber plants.

The publication *Diagnosis of mineral disorders in plants* (MAFF/ARC 1983) gives colour prints of mineral deficiency symptoms in vegetable crops (vol. 2) and protected crops (vol. 3). Colour prints of mineral deficiency symptoms in arboriculture plants are given in *Gebreksziekten in boomkwekerijgewassen* (Aendekerk).

7.14 FOLIAR SPRAYS

Applying microelements by foliar sprays (Table 7.12) is sometimes useful as a quick diagnostic aid, or as a means of correcting deficiencies when elements are 'fixed' or unavailable in the growing media. However, the leaves of some plants do not absorb sprays efficiently because of the wax in their cuticles, a surfactant is therefore required. It is usually safer to apply the sprays under cloudy conditions, iron-deficient leaves, for example, can be badly scorched if sprayed in strong sunlight. Elements such as iron and boron are not mobile in the plant and more than one spray may be required. Wherever possible the deficiency should be corrected in the growing medium.

Table 7.12 Microelements that can be applied as foliar sprays.

Element	Salt	Rate $(g\,l^{-1})$	Remarks
boron	borax	1	do not apply this solution to the growing medium
copper	copper oxychloride	1	better than copper sulphate
manganese	manganese sulphate	1	for chrysanthemums
		4	for tomato plants spraying is more effective than soil drench
molybdenum	sodium molybdate	0.5	deficiency soon shows with lettuce seedlings, apply $5\,g\,m^{-3}$ with base fertilizer; include 0.1 p.p.m. in liquid feed for *Poinsettia*
iron	Fe-EDTA	0.2–0.5	EDTA is safer for foliar spraying than EDDHA
	ammonium ferric citrate (16–22% Fe)	0.05	spray on cool, cloudy days

CHAPTER EIGHT

Mix formulation and preparation

8.1 HISTORICAL

A brief historical review of the development of loamless mixes shows the difficulty of attempting to establish a date for their first introduction or general acceptance into horticulture. As with many horticultural practices, evolution has occurred over a long period. Peat, leafmould and pine needles have been used for growing azaleas by many generations of gardeners, and experiments made at Versailles in 1892 on the nutrition of azaleas in these media are described by Watson (1913). Laurie (1931) in Ohio, USA, experimented in the late 1920s and early 1930s on the use of peat–sand mixtures for growing a range of plants, but no further development of this concept appears to have occurred for the next 20 years. In the 1950s, American workers at Michigan (Asen & Wildon 1953) and in California (Baker 1957) revived the interest in the use of peat and sand by obtaining favourable results in comparison with plants grown in traditional mixes based on mineral soils. In Europe, Penningsfeld, working in Bavaria, initiated work on the use of pure peat as a medium (Penningsfeld 1962), and Puustjärvi (1969), working in Finland, developed a system of growing in peat known as 'basin culture'. With this system, vegetable and ornamental crops are grown in peat, isolated from the glasshouse border soil by a sheet of polythene. Several attempts have been made to include clay, in either powdered or granular form, into mixes made from peat and sand. Fruhstorfer (1952) used a 1:1 mixture of granulated clay and peat to formulate his *'Einheitserde'* compost. Dempster (1958) recommended 10% by volume of clay to 40% peat and 50% sand, and Dänhardt & Kühle (1959) reported a progressive increase in growth as the amount of clay mixed with the peat was decreased to 10%; the inferior result when growing plants entirely in peat was attributed to the very low pH in the absence of some clay. Difficulties over supply and quality control have restricted the use of clay in mixes; early unpublished work by the author showed that adverse effects on pH, microelement availability and forms of nitrogen could occur if the clay contained more than a very small amount of calcium carbonate.

During the last three decades, there has been a large increase in the interest shown in loamless mixes, by research workers and growers alike; the main reasons for this were discussed in Chapter 1. Each issue of the abstracting journals covering the world's literature on horticulture and plant nutrition contains several references to the use of loamless mixes. These often differ considerably in their formulation, because of the wide variations in the characteristics of the materials used and differences in the nutritional requirements of various crops. For example, polystyrene does not contain any plant nutrients, whereas vermiculite is rich in magnesium and potassium; sedge peats have more available nitrogen than sphagnum peats etc. Also, a few plant species show distinct nutritional preferences, e.g. azaleas prefer ammonium to nitrate nitrogen, ericaceous plants prefer a lower pH than most other plants, and tomatoes require a high potassium level. Climatic conditions also alter the optional balance and concentration of nutrients.

In early work by the author on the nutrition of plants in peat–sand mixes, large-scale factorial experiments were used to study the simultaneous variation in the concentrations of a number of nutrients. For example, a 3^4 design giving a total of 81 treatments was used to study 3 levels each of nitrogen, phosphorus, potassium and calcium. A 2^6 design was used to study two levels of six factors, e.g. 3 factors being physical, such as ratios of peat to sand, grades of peat and grades of sand, with the other 3 factors being nutritional, such as amounts of microelements, sources of phosphorus and sources of nitrogen. This made a total of 64 treatments. These experiments were repeated in summer and winter to include the effect of the season, and it soon became clear that no one formulation was optimal for all plants under all conditions.

The very large number of published formulae precludes any attempt at compiling a comprehensive list, rather a selection has been made of the more widely known pot-plant mixes used in various countries.

8.2   DENMARK

A potting mix known as *Garta Jord* is marketed by the Co-operative Horticultural Supply Company. It is based on a mixture of dark peat, straw that has been composted with added nitrogen, and 'Grodan', a proprietary product made from a glasswood fibre produced from basalt rock and limestone. Its formulation is:

> 50% dark peat
> 25% composted straw
> 25% Blue Grodan

To this is added:

m^{-3}		yd^{-3}
9 kg	ground limestone	15 lb 3 oz
400 g	superphosphate	10 oz
200 g	potassium sulphate	5 oz
60 g	magnesium sulphate	2 oz
50 g	iron sulphate	1 oz
16 g	copper sulphate	0.4 oz
10 g	manganese sulphate	0.3 oz
8 g	zinc sulphate	0.2 oz
8 g	borax	0.2 oz
2 g	sodium molybdate	0.1 oz

This supplies the following amounts of macronutrients ($mg\,l^{-1}$):

N	P	K
—	32	88

No nitrogen is included in the base fertilizer; some nitrogen has already been added to assist in the decomposition of the straw and will be present in the mix largely in organic form. Liquid feeding with 0.2–0.5 $g\,l^{-1}$ of calcium or potassium nitrate is given shortly after potting.

The German 'Einheitserde T' potting mix, based on 50% of an acid montmorillonite clay and 50% peat, is also widely used in Denmark.

8.3 FINLAND

Professor Puustjärvi of the Peat Research Institute at Hyrylä has worked extensively on peat as a growing medium for bed- or border-grown crops such as tomato, carnation, etc., and has introduced the 'basin systems' of cultivation. This consists of sheets of polythene spread on top of the glasshouse soil; the polythene is turned up 1–2 cm at the sides and edges, and peat is placed on top to a depth of 10–30 cm, depending upon the type of crop. The polythene forms a shallow basin and prevents extensive loss of nutrients by leaching. For growing plants in pots, the following nutrient additions per cubic metre of sphagnum peat are recommended:

Dolomite limestone 8 kg
compound fertilizer 1–1.6 kg (1 lb 11 oz–2 lb 11 oz yd^{-3})

depending upon the vigour of the plant.

The fertilizer has an analysis of 11% N, 24% P_2O_5, 22% K_2O; at 1 $kg\,m^{-3}$ the nutrients supplied ($mg\,l^{-1}$) would be:

N	P	K
110	105	182

In addition, microelements included in the fertilizer would supply:

	B	Cu	Mn	Zn	Fe	Mo
mg l^{-1}	1	11	9	7	9	1

In the original formulation, several straight or single nutrient fertilizers were used, but a compound fertilizer is now preferred for simplicity. Normally, slow-release forms of nitrogen are not used; liquid feeding with frequent chemical analysis to check the level of plant nutrients is preferred.

8.4 GERMANY

Professor Penningsfeld from the Institute for Soil Science and Plant Nutrition, Weihenstephan, Nr Munich, has made large contributions to our knowledge of the nutrition of ornamental plants grown in peat. He has recommended three rates of nutrients, depending upon the vigour of the plants.

For all mixes, calcium carbonate is added at 2–5 kg m^{-3} of peat, the actual rate depending upon the lime requirement of the peat and the desired pH. A compound fertilizer having an analysis of 12% N, 12% P_2O_5, 16% K_2O, 2% MgO is then added at the following rates:

Medium 1. 0.5–1.0 kg m^{-3}, giving the following concentrations of applied nutrients:

	N	P	K
mg l^{-1}	60–120	26–53	66–133

This mix is used for growing salt sensitive plants: *Adiantum, Erica gracilis, Primula, Gardenia, Camellia,* and certain bedding plants, i.e. *Begonia, Verbena, Godetia, Callistephus* and *Dianthus.*

Medium 2. 1.5 kg m^{-3} of the compound fertilizer, giving:

	N	P	K
mg l^{-1}	180	80	200

This mix is used for growing moderately salt-tolerant plants: *Aechmea, Vriesia, Freesia, Anthurium, Gerbera, Aphelandra, Cyclamen, Monstera, Sansevieria, Rosa, Hydrangea,* and certain bedding plants, i.e. *Salpiglossis, Tagetes, Zinnia, Matricaria, Penstemon, Dianthus, Campanula.*

Medium 3. 3 kg m^{-3} of the compound fertilizer, giving:

	N	P	K
mg l^{-1}	360	160	400

This mix is used for growing plants that have the greatest salt tolerance:

Pelargonium, Euphorbia, Saintpaulia, Chrysanthemum, Asparagus and *Carnation.*

Microelements are added to these mixes in either liquid or metal alloy formulations. An example of a fertilizer supplying microelements in liquid form is 'Micro T', made by Gabi (Hundersen, 4902 Bad Salzuflen 1). This contains 0.02% B, 0.13% Cu, 0.05% Mo, 0.17% chelated iron, 0.05% Mn and 0.007% Zn and is applied at the rate of 1 litre m^{-3}. The microelements in this fertilizer are immediately available to plants. An alternative fertilizer used to supply microelements in a slow-release form is 'Radigen', manufactured by Metalldunger Jost (0–5860 Iserlohn Postfach 224). This fertilizer contains 2% Mg, 2% Fe, 1.5% Cu, 0.8% Mo, 0.8% Mn, 0.8% B and 0.6% Zn. It is based on metal alloys which are slowly attacked by the weak acids in the mix, thereby releasing the microelements over a long period, and is used at the rate of 100 $g\,m^{-3}$.

8.5 IRELAND

Workers at the Agricultural Institute, Kinsealy, Dublin, have developed a mix based on peat which is suitable for a wide range of plants and is known as the 'Range Mix'. The following fertilizers are added:

$kg\,m^{-3}$		yd^{-3}
1.4	calcium ammonium nitrate (26% nitrogen)	2 lb 4 oz
1.4	superphosphate	2 lb 4 oz
0.7	potassium sulphate	1 lb 2 oz
9.0	Dolomite limestone	15 lb
0.4	frit 253A	10 oz

This gives an added nutrient concentration ($mg\,l^{-1}$) of:

N	P	K
364	110	310

A mix designed specifically for tomato propagation contains the same amounts of Dolomite limestone and potassium sulphate as the Range Mix, but the calcium ammonium nitrate is reduced to 0.7 kg m^{-3} (18 oz yd^{-3}) and urea formaldehyde is added at 0.7 kg m^{-3} (18 oz yd^{-3}) to give some slow-release form of nitrogen.

8.6 NETHERLANDS

A standardized potting mix was introduced in the Netherlands in 1964 by the Aalsmeer Research Station for Floriculture (Arnold Bik, personal communication). It was known as the *Regeling Handelspotgrand Proefstation Aalsmeer*, abbreviated to RHPA (Regulation System for Commercial Potting Composts Research Station Aalsmeer). The present RHP system offers growers a range of mixes with various nutrient levels and physical properties. A general potting mix is made from equal volumes of young, white, fibrous, sphagnum peat and frosted, decomposed, black sphagnum peat. To one cubic metre is added: 6 kg 'Dolokal', a magnesium limestone (10% MgO), giving a pH of 5.5–6.0, and a compound fertilizer 'PG mix', having 14% N, 16% P_2O_5, 18% K_2O, the trace elements are supplied as inorganic salts except the iron, which is DTPA. The fertilizer rate for most bedding and pot plants would be 1.5 kg m^{-3}.

This gives an added nutrient concentration (mg l^{-1}) of:

N	P	K
210	105	224

and

B	Cu	Mn	Fe	Mo	Zn
0.45	1.8	2.4	1.35	3.0	0.6

The nutrient levels can be changed for specific crops. For example, for seed and potting mixes the rate of PG mix is reduced to 0.5 kg m^{-3}. For plants requiring a more acid mix, e.g. *calceolaria*, the pH is reduced to 5.0–5.5. The physical properties can also be varied, e.g. for blocking mixes the proportion of black peat is increased, and where irrigation is by 'ebb and flow' systems more fibrous peats are used.

8.7 NORWAY

Scientists at Kvithamar Experimental Station and the Department of Floriculture and Greenhouse Crops, Agricultural University, have developed a mix based on peat which is used for a wide range of crops:

kg m^{-3}		yd^{-3}
2.0	compound fertilizer (13–6–16)	3 lb 9 oz
3.0	rock phosphate	5 lb
3.5	ground limestone	5 lb 14 oz
1.5	Dolomite limestone	2 lb 8 oz
0.2	frit 36	13 oz

The added nutrient concentration (mg l^{-1}) is:

N	P	K
252	110 (soluble)	360
	450 (insoluble)	

The frit rate is increased to 0.5 kg m^{-3} for long-term crops or if the volume of growing medium per plant is small. The rock phosphate provides a reserve or long-term supply of phosphorus.

8.8   UNITED KINGDOM

Various nutritional problems arising from the use of peat-based mixes have been investigated by the author at the Glasshouse Crops Research Institute. Special attention has been given to problems in nitrogen and microelement nutrition, which occur most frequently. Results from this work have been used to formulate seed sowing and potting mixes.

GCRI Seed Mix

50% by volume sphagnum peat
50% by volume fine, lime-free sand

To this is added:

kg m^{-3}		yd^{-3}
0.75	superphosphate	1 lb 4 oz
0.4	potassium nitrate	10 oz
3.0	ground limestone	5 lb 4 oz

The high sand content of the seed sowing mix is to meet some growers' preference for media that allow rapid removal of the seedlings with the minimum of root disturbance. This mix is low in nutrients and is suitable only for seed germination; seedlings should be pricked-out as soon as possible after germination, otherwise they will starve. If for some reason pricking-out is delayed, a solution of calcium nitrate at 1.5 g l^{-1} (0.2 oz gal^{-1}) will maintain the growth of the seedlings until they can be handled. A fritted trace element mixture, WM 255, used at 375 g m^{-3} when sowing lettuce seeds will prevent molybdenum deficiency arising in the seedling stage. Normally, microelements are not required in seed sowing mixes.

GCRI Potting Mix   A general purpose potting mix is made from:

75% by volume sphagnum peat
25% by volume mineral aggregate

The type of aggregate and its particle size distribution should be chosen with respect to the grade of peat. If the peat has fine particles and little fibre, and a mechanical mixer is to be used, the aggregate particles should be coarse, i.e. grading 1–3 mm. Lightweight aggregates, such as perlite, do not increase the bulk density of the mix as do sand or grit. To this mixture is added:

$kg\,m^{-3}$		yd^{-3}
0.4	ammonium nitrate	10 oz
1.5	superphosphate	2 lb 10 oz
0.75	potassium nitrate	1 lb 5 oz
2.25	ground limestone	4 lb
2.25	Dolomite limestone	4 lb
0.375	Fritted trace elements (WM 255)	10 oz

This gives an added nutrient content $(mg\,l^{-1})$ of:

N	P	K
230	120	290

Approximately 30% of the nitrogen is in the ammonium form and 70% in the nitrate form.

This mix can be used immediately or stored for an indefinite period before use. If an all-peat mix is made using a young fibrous peat, the fertilizer remains the same apart from the liming materials which must be increased to give a pH of 5.0–5.5.

GCRI Potting Mix II   If a mix with a slow-release form of nitrogen is required, the ammonium nitrate in the above formulation can be omitted and urea-formaldehyde used at rates ranging from 0.5–1.0 kg m^{-3} (1–2 lb yd^{-3}) depending upon the vigour of the species and the season. Slow growing plants, in winter, would receive the low rate and vigorous plants, in summer, the high rate. *Do not store mixes having more than 0.5 kg m^{-3} of urea-formaldehyde.*

GCRI Potting Mix III   If there is the risk of phosphorus deficiency developing in the mix and phosphorus cannot be included in the liquid feed, the superphosphate in potting mix I can be omitted, the ammonium nitrate reduced to 0.2 kg m^{-3} (5 oz yd^{-3}), the potassium nitrate reduced to 0.4 kg m^{-3} (10 oz yd^{-3}) and one of the magnesium-ammonium-phosphate fertilizers, such as 'MagAmp' or 'Enmag', used at the rate of 1.5 kg m^{-3} (2 lb 8 oz yd^{-3}). This mix supplies sufficient phosphorus to grow a ten-week pot chrysanthemum crop without the need to include phosphorus in the liquid feed. The phosphorus supply would not, however, be adequate for plants grown for a longer period, e.g. *Cyclamen*.

In all the above mixes, the quantity of ground limestone and Dolomite limestone is based upon the lime requirement of Irish sphagnum peat. If peats with a different lime requirement are used (§ 4.7), the amount of liming materials must be varied accordingly. For *Erica*, the pH of the mix should be 4.0–4.5, measured in a suspension with a 0.01 M $CaCl_2$. The posssible rise in the pH of the mix due to 'hard' water must be borne in mind also. Plants such as *Cineraria*, *Calceolaria* and *Primula* have been found to grow better when the pH of the mix is reduced to 4.5–5.0.

Frit WM 255 with the higher molybdenum content is recommended for lettuce, *Brassica* and *Poinsettia*. Frit 253A with the high boron content is sometimes preferred for tomato propagation.

8.9   UNITED STATES OF AMERICA

Two groups of mixes are widely known and used in the USA.

8.9.1   *The University of California system*

Professor Baker and his colleagues at the University of California have introduced the UC System of growing. In this system, mixes can be made entirely of sand, or peat and sand, or of peat only. There are five different physical mixes and to each of these there are six different base fertilizers, making a total of thirty different mixes. In practice only a few of these are used and the following are among the most useful:

UC Mix D

75% by volume peat moss
25% by volume fine sand

To this mixture is added either of the following fertilizers:

Fertilizer I(D): this contains only a low amount of inorganic nitrogen.

$kg\,m^{-3}$		yd^{-3}
0.15	potassium nitrate	4 oz
0.15	potassium sulphate	4 oz
1.2	superphosphate	2 lb
3.0	Dolomite limestone	5 lb
2.4	calcium carbonate	4 lb

This gives an added nutrient concentration ($mg\,l^{-1}$) of:

N	P	K
20	95	123

An alternative base fertilizer mixture which contains some nitrogen in a reserve form as well as some inorganic nitrogen is:

Fertilizer II(D):

$kg\,m^{-3}$		yd^{-3}
1.5	hoof and horn	2 lb 8 oz
0.15	potassium nitrate	4 oz
0.15	potassium sulphate	4 oz
1.2	superphosphate	2 lb
3.0	Dolomite limestone	5 lb
3.0	calcium carbonate	4 lb

These rates of fertilizers supply:

	N	P	K
mineral $(mg\,l^{-1})$	20	95	123
organic $(mg\,l^{-1})$	195		

UC Mix E   This is an all-peat mix and, again, fertilizers can be used at two rates.

Fertilizer I(E): this contains inorganic nitrogen only.

$kg\,m^{-3}$		yd^{-3}
0.2	potassium nitrate	6 oz
0.6	superphosphate	1 lb
1.5	Dolomite limestone	2 lb 8 oz
3.0	calcium carbonate	5 lb

The added nutrient concentration $(mg\,l^{-1})$ is:

N	P	K
26	48	76

Fertilizer II(E): this contains some reserve or organic nitrogen as well as inorganic nitrogen.

$kg\,m^{-3}$		yd^{-3}
1.5	hoof and horn	2 lb 8 oz
0.2	potassium nitrate	6 oz
0.6	superphosphate	1 lb
1.5	Dolomite limestone	2 lb 8 oz
3.0	calcium carbonate	5 lb

This supplies:

	N	P	K
mineral (mg l^{-1})	26	48	76
organic (mg l^{-1})	195		

No microelement additions are recommended for these mixes. An interesting feature of these formulations is the reduction in the amount of liming materials as the sand is decreased and the peat increased. It appears that the fine sand in California is often slightly acidic in reaction; the rate of superphosphate is also reduced as the sand content of the mix is decreased. Steam sterilization of the mix is generally recommended.

8.9.2 Cornell Peat–lite mixes

The second group of widely known loamless mixes in the USA is the Cornell Peat–lite mixes. These mixes were introduced by Professors Boodley and Sheldrake of Cornell University, New York, and are based on mixtures of peat and vermiculite, or peat and perlite. The fertilizer levels are varied slightly, depending upon the type of plant to be grown and whether liquid feeding or slow-release fertilizers are to be used.

Peat–lite Mix A for pot plants comprises:

> 50% by volume sphagnum peat
> 50% by volume vermiculite

To this is added:

kg m^{-3}		yd^{-3}
0.9	calcium or potassium nitrate	1 lb 8 oz
0.6	superphosphate	1 lb
3.0	ground limestone	5 lb
0.07	fritted trace elements (FTE no. 503)	2 oz

This supplies (mg l^{-1}):

N	P	K
117	48	340

Peat–lite Mix B. The vermiculite is replaced with perlite:

> 50% by volume sphagnum peat
> 50% by volume perlite

The fertilizers are the same as the Mix A, except that potassium nitrate

184

must be used and not calcium nitrate; this is because the vermiculite contains available potassium whereas the perlite does not. The amount of superphosphate added must also be doubled. The fritted trace element mixture has an analysis of:

	B	Cu	Fe	Mn	Zn	Mo
%	3	3	18	7.5	17	0.2

With these two mixes, no slow-release fertilizers are added and liquid feeding would start 2–3 weeks after potting. If a complete NPK slow-release fertilizer is required, Osmocote 14–14–14 is used at the rate of $3 \, kg \, m^{-3}$ ($5 \, lb \, yd^{-3}$).

A feature of the Cornell Peat-lite mixes is the use of a non-ionic wetting agent, $85 \, cm^3$ being added to 30–60 litres of water for each cubic metre of mix (3 fl oz to 5–10 gal yd^{-3}). Alternatively, the wetting agent can be added first to a small amount of vermiculite, which is then added to the bulk mix.

Recently, two additional mixes have been introduced for foliage and epiphytic plants.

Cornell Foliage Plant Mix is recommended for plants which require a mix with high moisture retention characteristics. The bulk mix comprises:

> 2 parts by volume sphagnum peat
> 1 part by volume vermiculite (grade no. 2)
> 1 part by volume perlite (medium grade)

To this is added:

$kg \, m^{-3}$		yd^{-3}
0.6	potassium nitrate	1 lb
1.2	superphosphate	2 lb
1.6	10–10–10 fertilizer	2 lb 12 oz
4.9	Dolomite limestone	8 lb 4 oz
0.4	iron sulphate	12 oz
0.07	fritted trace elements	2 oz

This supplies the following nutrients:

	N	P	K
$mg \, l^{-1}$	238	165	360

This mix is recommended for such plants as: *Aphelandra squarrosa, Begonia, Beloperone guttata, Cissus, Coleus,* Ferns, *Ficus, Hedera, Maranta, Pelargonium, Pilea, Sansevieria.*

Cornell Epiphyte Mix is used for plants that require good drainage and aeration and are able to withstand drying out between waterings. The bulk mix comprises:

1/3 by volume sphagnum peat (screened $\frac{1}{2}$ in mesh)
1/3 by volume Douglas red or white fir bark ($\frac{1}{8}$ in–$\frac{1}{4}$ in)
1/3 by volume perlite (medium grade)

To this is added:

$kg\,m^{-3}$		yd^{-3}
0.6	potassium nitrate	1 lb
2.7	superphosphate	4 lb 8 oz
1.5	10–10–10 fertilizer	2 lb 8 oz
4.2	Dolomite limestone	7 lb
0.3	ferrous sulphate	8 oz
0.07	fritted trace elements	2 oz

This gives an added nutrient concentration of:

	N	P	K
$mg\,l^{-1}$	228	280	352

Plants grown in this mix are: bromeliads, cacti, *Crassula*, *Diffenbachia*, *Episcia*, *Gloxinia*, *Hoya*, *Monstera*, *Philodendron*, *Peperomia*.

If a mix with slow-release nutrients is required, the fertilizers other than Dolomite limestone and superphosphate are omitted and either Osmocote 14–14–14 or Peter's 14–7–7 added at 3 kg m^{-3} (5 lb yd^{-3}).

8.9.3 Pennsylvania State peat formulae

Two mixes for propagating and growing ornamentals have been introduced by the Pennsylvania State University (White 1974). They are both based on sphagnum peat.

The Propagation Mix contains:

$kg\,m^{-3}$		yd^{-3}
0.9	superphosphate	1 lb 8 oz
0.3	potassium nitrate	8 oz
0.15	complete fertilizer (20–19–18)	4 oz
3.0	Dolomite limestone	5 lb

This gives a nutrient concentration of:

	N	P	K
$mg\ l^{-1}$	69	83	136

The Growing-on Mix contains:

$kg\ m^{-3}$		yd^{-3}
1.2	superphosphate	2
0.6	potassium nitrate	1
0.6	complete fertilizer (20–19–18)	1
3.0	Dolomite limestone	5

The nutrient concentrations are:

	N	P	K
$mg\ l^{-1}$	192	143	317

The $20N–19P_2O_5–18K_2O$ fertilizer is water-soluble and can also be used to make up liquid feeds. It contains microelements, mainly in chelated form.

8.10   HARDY NURSERY STOCK

In the UK this has now become the most important sector in terms of the volume of media used. In 1984 there were 76 million plants grown, the average size of container being 3 litres. Although the general physical and chemical principles relating to container mixes for nursery stock are similar to those for vegetables and ornamentals grown under protected cropping, there are certain major differences with respect to crop environment, management and specific nutritional requirements that must be recognized. These include the following:

(a) Usually a greater bulk density is necessary to give stability to tall plants, especially on windswept sites. Adding 15 or 25% by volume of grit increases the bulk density of a sphagnum peat mix from approximately 0.1 g cm^{-3} to 0.25 and 0.45 g cm^{-3} respectively. Traditional loam-based mixes have a bulk density of about 1 g cm^{-3}.

(b) In winter, when evapotranspiration is low, rainfall is often high, and if the mix is not well-drained root damage and plant death may result. Aeration is improved by using a coarse, fibrous peat or by adding coarse aggregates to the mix, e.g. grit or bark chips; the containers can also be stood on sand beds. As well as improving aeration by removing excess water in winter, containers can also be irrigated in this way in summer (§ 10.2.1).

187

(c) Container plants can require up to 2.5 cm (1 in) of irrigation water per day in some environments. To reduce the potentially large loss of nutrients that can occur by leaching, or their inefficient application by overhead spray irrigation, controlled-release fertilizers such as Osmocote, Nutricote or Ficote are used almost universally. Fertilizers that are temperature dependent for their nutrient release also have advantages where crops are overwintered outside.

(d) Apart from varying the rates of fertilizer in relation to plant vigour and container size, two special nutritional categories are recognized. Ericaceous plants require a low pH, and phosphorus-sensitive species require lower than normal levels of soluble phosphorus.

8.10.1 Phosphorus toxicity

General aspects of phosphorus toxicity were discussed in Chapter 6. Certain hardy nursery stock species are susceptible to high levels of soluble phosphate fertilizer application, especially when grown in loamless media that do not fix phosphorus; about 60% of the superphosphate added to these media remains in water-soluble form. The occurrence of phosphorus toxicity is also dependent upon the balance between levels of available phosphorus and other elements. Increasing the levels of nitrogen and potassium will often reduce the symptoms of phosphorus toxicity. Reported plant symptoms associated with high levels of soluble phosphorus have included: pale leaves, leading to chlorosis and iron deficiency symptoms; chlorosis and tip die-back in *Cytisus*; bronzing and leaf drop in *Elaeagnus pungens* 'Maculata'. *Camellia* grown in high

Table 8.1 Effects of superphosphate rates and slow-release fertilizer sources on the percentage of phosphorus in the tissue and the total plant dry weight of three species of ornamental plants. Bare-rooted cuttings were potted April 1981 and sampled Dec–Jan 1982. All controlled-release fertilizers were given at a rate of 3 kg m^{-3}

Slow-release fertilizer	Rate of superphosphate (kg m^{-3})	*Juniperus communis* 'Hibernica'		*Cytisus* x *praecox*		*Viburnum burkwoodii*	
		% P	DW (g)	% P	DW (g)	% P	DW (g)
Osmocote	0	0.24	18	0.50	18	0.31	14
18–11–10	0.75	0.21	20	0.60	9	0.38	13
(8–9 months)	1.5	0.24	21	0.71	11	0.38	10
Nutricote	0	0.20	9	0.15	19	0.15	8
16–10–10	0.75	0.21	21	0.32	19	0.36	26
(140 days)	1.5	0.23	24	0.63	12	0.37	18
Ficote	0	0.21	17	0.22	35	0.14	19
14–14–14	0.75	0.21	27	0.73	21	0.33	25
(70 days)	1.5	0.23	23	0.75	9	0.46	19

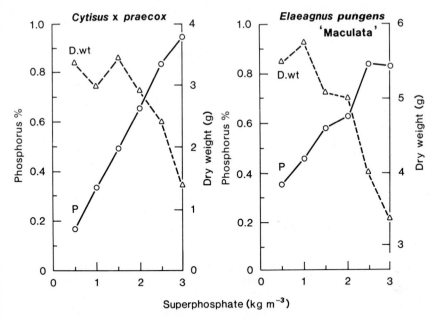

Figure 8.1 Phosphorus toxicity in two hardy ornamentals grown in a peat–sand mix. Increasing rates of superphosphate increased the percentage of phosphorus in the tissue but reduced the dry weight of the plants (from Scott 1982).

phosphate mixes have been more prone to the fungal disease *Monochaetia karstenii* (*Pestalotiopsis guepini*).

Some effects of the rate of superphosphate in mixes on the percentage of phosphorus in the leaf tissue, and the plant dry weight, are shown in Figure 8.1. Interactions between the rates of superphosphate used in the base fertilizer, the type of controlled-release fertilizer and the plant species are shown in Table 8.1. When *Juniperus* and *Viburnum* were grown with Nutricote and Ficote, there was a growth response to superphosphate at 0.75 kg m^{-3}; however, with Osmocote there was no response, an adequate amount of phosphorus having been released from the fertilizer. The following are examples of mixes used in various countries to grow hardy ornamental nursery stock in containers.

8.10.2 Efford Experimental Station, UK, nursery stock mixes

These mixes are based on 7 parts peat, 3 parts coarse chip bark and 2 parts coarse, lime-free grit, i.e. peat 58.3%, bark 25%, grit 16.7%. If the bark has not been composted with added nitrogen, it is recommended that for each 10% of bark in a cubic metre of mix, 100 g of ammonium nitrate be added in addition to the fertilizers listed below to compensate for the lock-up of nitrogen. It will also be necessary to reduce the quantity of liming materials as the proportion of bark is increased.

Lime and trace elements   The amounts of liming materials are adjusted for the quality of the water and whether or not the plants are calcifuge. Calcifuge species include: *Azalea, Calluna, Camellia, Elaeagnus, Erica, Magnolia, Pieris, Rhododendron, Skimmia* and most conifers.

Water quality	Rates of magnesium limestone ($kg\,m^{-3}$)	
	general species	calcifuge species
>200 mg l^{-1} bicarbonate	2.4	1.2
<200 mg l^{-1} bicarbonate	2.4*	2.4
	Fritted trace elements ($kg\,m^{-3}$)	
	general species	calcifuge species
235A or 255	0.3	0.3

* Addition of 1.2 kg m^{-3} calcitic limestone or chalk required.

Nutrition   Other nutrients are supplied by controlled-release fertilizers, either in 8–9 month formulations for one season's growing without liquid feeding (e.g. Osmocote 18–11–10 and Ficote 70 16–10–10) or in 12–14 months formulation for spring sales or where plants will be grown for a second season without potting on (e.g. Osmocote 17–10–10 and Ficote 140 16–10–10). As the nutrient release rates from these fertilizers is controlled by temperature, significant deviations in either seasonal temperatures or rainfall may necessitate supplementary liquid feeding.

	Rates of controlled-release fertilizers ($kg\,m^{-3}$)			
	single season (8–9 months)		extended season (12–14 months)	
	moderate or vigorous	slow growing or salt sensitive	moderate or vigorous	slow growing or salt sensitive
Liner mix (rooted cuttings into 7–9-cm pots)				
autumn potting	1.5	0.75	2.0	1.5
spring potting	3.0	1.5	3.0	2.5
Potting Mix (rooted cuttings into 2–3-litre pots, ericaceous into 0.4–1.0-litre pots)	3.0	1.5	4.0	3.0

The above rates are for containers in the open with overhead irrigation; when the containers are sub-irrigated or are under polythene, use only 2/3 of these rates of fertilizers.

No superphosphate is required when using Osmocote, with Ficote 16–10–10 use 0.75 kg m^{-3} of single superphosphate. If straight rather than controlled-release fertilizers are used, single superphosphate is included in the base fertilizer at 1.5 kg m^{-3}, for Ericaceous and phosphorus-sensitive plants reduce this to 1 kg m^{-3}. Species found to be sensitive of high levels of phosphorus include:

evergreen *Azalea*	*Hydrangea hortensia*
deciduous *Azalea*	*Senecio greyi* 'Sunshine'
Camellia japonica	*Viburnum burkwoodii*
Magnolia soulangiana	*Chamaecyparis lawsoniana*
Elaeagnus pungens 'Maculata'	'Ellwoods Gold'
Skimmia japonica 'Foremanii'	*Chamaecyparis pisifera*
Cytisus x *praecox*	'Boulevard'

8.10.3   Oklahoma State University mixes

Professor C. E. Witcomb and co-workers at Oklahoma State University recommend three fertilizer rates and combinations, depending upon the duration of the crop. The physical components are:

3 parts by volume ground pine bark
1 part by volume peat
1 part by volume sand

Early recommendations had included phosphorus as either single or triple superphosphate, subsequent work has shown this to be unnecessary.

	kg m^{-3}	yd^{-3}
For all mixes include:		
Dolomite limestone	3.6	6 lb
Micromax (trace element mix)	0.9	1 lb 8 oz
For 9–12 month crops use:		
Osmocote 17–7–12	5.9	10 lb
Osmocote 18–6–12	2.4	4 lb
For medium crops, 5–6 months, use:		
Osmocote 18–6–12	7.1	12 lb
For short-term crops, 3–4 months, use:		
Osmocote 18–6–12	3.6	6 lb
Osmocote 19–6–12	2.4	4 lb

Osmocote	Formulation (months)
17–7–12	12–14
18–6–12	6–9
19–6–12	3–4

The essential difference between these fertilizers is their rates of release, not their nutrient contents. Mixtures of two fertilizers are used to achieve the desired rates of release. Much higher rates of controlled-release fertilizers are used in these mixes than in the UK because of the very high rates of irrigation (approximately 2.5 cm (1 in) per day during mid-summer).

8.10.4   New Zealand

Dr M. Prasad, of the Levin Horticultural Research Centre, recommends the following fertilizer rates for container-grown nursery stock in a pine bark mix:

$kg\,m^{-3}$		yd^{-3}
1	superphosphate	1 lb 11 oz
1	calcium ammonium nitrate	1 lb 11 oz
3	Osmocote 18–11–10 (9 months)	5 lb
4–5	Dolomite limestone	6–8½ lb
	trace element mixture	

Osmocote 18–11–10 is also known as 18–4.8–8.3, i.e. the element rather than the oxide basis. It can be used at 5 kg m^{-3} for vigorous plants.

The trace element mixture consists of: 11.8 g m^{-3} borax, 21.2 g m^{-3} copper sulphate, 35.4 g m^{-3} ferrous sulphate, 50 g m^{-3} chelated iron, 14.2 g m^{-3} manganese sulphate, 14.2 g m^{-3} zinc sulphate and 2.4 g m^{-3} sodium molybdate.

The milled pine bark (*Pinus radiata*) has a particle size distribution:

100% <5 mm, 70–85% <2.5 mm, 30–60% <1 mm, 10–20% <0.5 mm

No phytotoxicity has been reported if the bark is stored in the open for 4–5 weeks after milling. It is not composted unless it has a high percentage of wood.

8.11   AZALEA MIXES

Azaleas will grow satisfactorily only in a mix which has a relatively low pH, i.e. in the range 4.5–5.5; also the salinity or nutrient levels must be lower than is customary for most plants. For these reasons, mixes which are used for the majority of pot plants are not acceptable for azalea growing. Nutritional studies made with plants grown in sand cultures have shown that the azalea grows best when most of the nitrogen is in the ammonium form. This form of nitrogen reduces the uptake of other cations and helps to maintain a low pH within the leaf tissue. The availability of iron is thereby increased and the incidence of iron

192

deficiency, a disorder to which azaleas are particularly susceptible, is reduced. The pH of the mix must also be kept low during the growing season. In areas having water with a high calcium content, some of the carbonates must be neutralized by treating the water with nitric acid, if phosphoric acid is used at high rates it can induce iron deficiency; alternatively, acidic liquid feeds, based largely on ammonium sulphate, can be used to prevent the rise in pH which would otherwise occur. It must also be remembered that, at the low pH values maintained in azalea mixes, boron toxicity can develop quickly if too much boron is present in either the water or the fertilizers used to make up the liquid feed. Azaleas are also very susceptible to salt injury arising from high levels of fertilizers. The injury is seen first as a yellowing of the leaves which later turn brown and drop off. Mixes having a poor physical structure with a low air-filled porosity, and which remain excessively wet after being irrigated, can also cause the leaves to drop. Aspects of the mineral nutrition of *Azalea* have been reviewed by Kofranek & Lunt (1975).

In Belgium and the Netherlands, the traditional material for growing azaleas has been pine needles. However, the rapid expansion in the number of azaleas grown has forced growers into using other media, principally sphagnum peat, sometimes with a quantity of shredded bark or foam plastic added to improve the aeration. These media have lower levels of available nutrients than pine needles and more attention must be given to the liquid feeding.

8.11.1 Belgium

Blomme & Piens (1969) analysed the media in which a large number of azaleas of different ages were growing, and related the analysis to the growth of the plants. They concluded that the most favourable levels of nutrients were: 11 mg of phosphorus, 33 mg of potassium and 357 mg of calcium per 100 g of medium. Magnesium levels did not appear to influence growth and no data were given for nitrogen levels.

Also in Belgium, Gabriels *et al.* (1972) determined the optimal balance of anions and cations for *Azalea* cv. 'Ambrosiana' grown in peat or in coniferous litter. Peat-grown plants were considered superior to those grown in the coniferous litter and the best results were obtained when a total of 2.08 g nitrogen, 0.35 g phosphorus, 2.72 g potassium, 1.61 g calcium, 0.98 g magnesium and 0.15 g sulphur was applied to each plant over the growing season in liquid feeds given at weekly intervals. From these data, the optimal ratio of the three main nutrients in the fertilizer can be calculated as $1 \, N : 0.17 \, P : 1.3 \, K$.

The analysis of the media at the end of the growing season was: pH 5.1 (in water), 4.5 (in 0.1 M KCl); conductivity $0.427 \, mS \, cm^{-1}$. The extractable nutrients were:

	NO_3	P	K	Ca	Mg
$mg \, l^{-1}$	201	38	190	655	270

Mineral composition of the leaf tissue was:

N	P	K	Ca	Mg	SO_4
2.6%	0.49%	2.0%	0.75%	0.28%	0.71%

8.11.2 Netherlands

Arnold Bik (1972) studied the nutrition of *Azalea* cv. 'Ambrosiana' during the period when the young plants are grown in beds out of doors, prior to being potted up and brought into the glasshouse for forcing, i.e. from the end of May to the end of September. The plants were grown in beds of decomposed sphagnum peat to which magnesium and microelements were added prior to planting. All other nutrients were applied in liquid feeds on three occasions during the season. It was concluded that the best results were obtained when the total amounts of nutrients applied during the season were:

	N	P	K
$g\ m^{-2}$ of bed	72	7.8	19.9

This is equivalent to a total of:

$194\ g\ m^{-2}$ ammonium nitrate
$37.4\ g\ m^{-2}$ monoammonium phosphate } supplied in the three liquid feeds
$48.5\ g\ m^{-2}$ potassium nitrate

The leaves of the plants receiving these rates of nutrients had the following mineral analysis:

N	P	K
2.19%	0.18%	0.70%

The organic salt content of the azalea leaves, defined as the sum of the cations K^+, Na^+, Ca^{2+} and Mg^{2+} minus the sum of the anions NO_3^-, $H_2PO_4^-$, SO_4^{2-} and Cl^-, was found to be only one half of the value found in chrysanthemum leaves and only one third of the value in gloxinia leaves, thereby showing a distinct ionic behaviour for this species.

8.11.3 United States of America

In the USA Batson (1972) recommended a coarse grade of sphagnum peat with a pH of 5.0 as the growing medium. No base fertilizer is mixed with the peat, apart from chalk if it is necessary. Feeding is commenced two weeks after potting and then repeated at weekly intervals, one of three formulations of liquid feed being used. The choice depends upon the pH of the peat, which is determined at weekly intervals.

If the pH is between 4.9 and 5.4, liquid Feed 'A' (23–10–12) is used:

g per 100 l		oz per 100 US gal
78.6	ammonium nitrate	10.5
45	diammonium phosphate	6
45	potassium nitrate	6

This solution has approximately:

N	P	K
430	100	170 p.p.m.

If the pH rises above 5.5, use Feed 'B' (21–0–0):

g per 100 l		oz per 100 US gal
180	ammonium sulphate	24
120	ferrous sulphate	16

This feed is acidifying and will reduce the pH; it has 378 p.p.m. nitrogen.
If the pH drops below 4.8, Feed 'C' (15–3–3) is used to raise the pH again:

g per 100 l		oz per 100 US gal
314	calcium nitrate	42
22.5	monocalcium phosphate	3
30	potassium nitrate	4

This feed contains:

N	P	K
511	54	114 p.p.m.

No feed must ever be given when the medium is dry; an irrigation with plain water should be given first. Feeding should be discontinued for seven days preceding each occasion when the plants are pinched or stopped, and should not be resumed until the new growth is 1.25 cm (0.5 in) long. The salinity of the media must be monitored weekly and, during the period of pinching, it should not exceed a value of 0.6 mS cm^{-1} on a Solu-Bridge test based on one part of medium to five parts of water; for a test based on one part of medium to two parts of water, the equivalent value would be 1.1 mS cm^{-1}. During the remainder of the growing period,

the salinity values should not exceed 0.90 on a 1:5 test or 1.9 on a 1:2 test. Regular feeding with iron in chelated form should be given at 30 g 100 l^{-1} (4 oz 100 gal^{-1}) every month and always with the first feed after every pinch. Nutrient levels should be determined each month and, on the spur-
way system of analysis, should be about 25–30 p.p.m. nitrogen, 3–5 p.p.m. phosphorus, 15–20 p.p.m. potassium and 100–150 p.p.m. calcium.

8.12   PROTEA MIX

Proteas are particularly sensitive to phosphorus toxicity (§ 6.1.4) and require very low levels of soluble phosphorus in the mix. The following mix has been used successfully in New Zealand (Prasad 1983) for *Leucodendron salignum* 'Red Bird' and 'Safari Sunset'. It is based on:

9 parts by volume fine pine bark
1 part by volume sterilized soil

To this is added:

kg m^{-3}		yd^{-3}
either		
0.8	ammonium sulphate	1 lb 6 oz
or		
0.5	calcium ammonium nitrate	13 oz
plus		
0.3	potassium sulphate	8 oz
0.2	magnesium sulphate	5 oz
1–2	Osmocote 19–6–10	1 lb 11 oz–3 lb 6 oz
	trace element mixture	

The trace element mixture is the same as that used for the nursery stock mix (§ 8.10.4). A liquid feed having 100 p.p.m. nitrogen and 83 p.p.m. potassium but without any phosphorus is used.

8.13   PROPRIETARY FORMULATIONS

The number of proprietary formulations has increased substantially during the last decade. In some situations proprietary mixes are more convenient for the grower and can also be as economical as own made mixes. Often their formulations are not disclosed, but in some countries manufacturers are obliged to state the amounts of added nutrients. Where the nutrient additions are not given, the experience of local users is usually the best guide. Boodley (1981) has compared the nutrient levels of 15 commercial mixes in the USA, and Bugbee & Frink (1983) made physical and chemical examinations of 54 commercial mixes.

If slow-release forms of nitrogen or controlled-release fertilizers have been used, and either the conditions or period of storage are unknown, there is the potential risk of the occurrence of either ammonia toxicity or high salinity (§ 4.8.2 and 5.4.1).

8.14   MIX PREPARATION

8.14.1   Fertilizer strength and balance

The various effects which the *concentration* of fertilizers have on plant growth, and the ways in which this can be manipulated by the grower, are often not appreciated. One effect of high rates of soluble fertilizers is to increase the osmotic potential of the medium which decreases growth, especially of seedlings and young plants. However, if the plants are not too severely damaged they will ultimately respond to the higher nutrient levels and make the most growth. An example of this effect is shown in Figure 8.2.

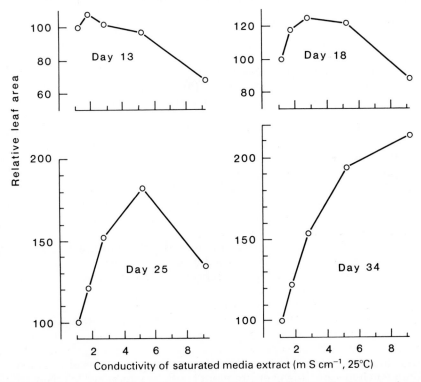

Figure 8.2   The effect which nutrient concentration has on the growth of tomato plants. With young seedlings there was a progressive reduction in growth with increasing amounts of nutrients; older plants responded to the higher rates of nutrients (Love & Bunt, unpublished).

The leaf areas of tomato plants grown with increasing rates of a balanced NPK fertilizer have been expressed relative to plants grown with the lowest rate. Initially, there was a progressive decrease in leaf area with increasing salinity. At days 18 and 25 the intermediate fertilizer levels were giving the most growth, and at day 34 there was a clear response to the highest fertilizer rate. The containers had been stood in saucers and there had been no loss of nutrients by leaching. The highest growth rates would have been achieved with a moderate rate of base fertilizer and early liquid feeding.

Other effects of high salinity which growers use to modify plant growth are: (a) a reduction in plant height; (b) darker foliage; (c) a temporary check to the vegetative growth of tomato plants, used to assist fruit setting under poor light conditions. The term 'osmotic growing' is sometimes used to describe these practices. The amounts of soluble fertilizers used in preparing mixes must, therefore, always be a compromise; the factors to consider being: season, species and management practices.

Lack of balance between the major nutrients can also depress plant growth. Sometimes increasing the concentration of a major element, even if it is not at deficiency level but is low in relation to other major elements, will improve plant growth.

8.14.2 Mixing

The main objectives for growers making their own mixes are to achieve: (a) physical and chemical uniformity, (b) consistency between mixes and (c) economy of labour. Attention should be given to the following points:

(a) By adding the bulk ingredients and fertilizers in a systematic manner there is less risk of either omissions or double applications.
(b) Powdered fertilizers should never be added to wet ingredients, even prolonged mixing fails to give chemical uniformity.
(c) If the mix requires wetting, water should be added through a spray or dribble bar *after* the fertilizers have been mixed. Most sphagnum peats should not be wetter than 60% on the moist weight basis before adding the fertilizers.
(d) Fertilizers such as superphosphate and potassium nitrate, that are liable to 'cake', should be crushed before weighing.
(e) Premixing the fertilizers with a carrier, e.g. dry sand, helps uniformity of distribution.

Mixing machines can be classified into two groups: continuous process and batch mixers.

Continuous process This method has much greater rates of production than batch mixers, however, frequent checks are required to ensure that the bulk ingredients and fertilizers are being used at the correct rates. Confirmation that the correct amounts have been used after a day's mixing does not constitute adequate quality control. With the continuous system of production, hoppers of separate ingredients are positioned in tandem over a moving belt. The correct ratio of each ingredient is discharged by an auger, moving belt or vibrator at the base of each hopper. A counter-rotating disc mixer positioned at the end of the belt gives partial mixing, which is completed by tumbling the mix onto a pile or an elevator.

Batch mixers   Several different principles are employed, some examples are shown in Figure 8.3. Not only must the mix be physically and chemically homogeneous, the structure should not be destroyed by too vigorous mixing. Mixers with a tumbling action are the least destructive of structure, whereas machines in which the rotor or agitator is in

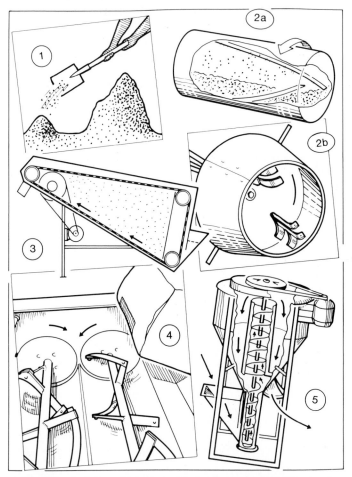

Figure 8.3   Methods of mixing with an increasingly vigorous action. The structure of some mixtures, e.g. peat–sand, can be more affected by the type of mixing action than others, e.g. peat–bark.

maximum contact with the stator or casing can be very destructive of structure and can also damage the coating of controlled release fertilizers. It is also important that the machine will discharge quickly, continued mixing during a slow discharge results in overmixing and a poor structure. The adverse effect which some mixers can have on structure can be seen from the following. A medium made from 75% sphagnum peat and 25% fine sand had an air-filled porosity of 6.9% when mixed by hand and only 2.9% when mixed in a horizontal twin-rotor mixer.

Quality control   Mixes should be visually checked to ensure there is neither segregation of ingredients nor that the structure is poor. That the moisture content is within an acceptable range can be checked by feel. Both the chemical uniformity within a batch and consistency between batches can be quickly monitored by conductivity tests.

CHAPTER NINE

Liquid feeding

When plants are grown for long periods in small volumes of media, the supply of available nutrients becomes depleted, growth ceases and symptoms of nutrient deficiency may spoil the general appearance of the plants. Years ago, the grower's remedy was either to occasionally add to each pot a teaspoonful of powdered fertilizer which was washed into the media at the next watering, or to make up a liquid feed by suspending animal manure in coarse sacking in a barrel of water and allowing the soluble nutrients to diffuse into the water. This solution was then applied to the pots with a watering can. These methods of liquid feeding are too imprecise and time consuming to be used today and the practice of adding soluble chemicals to the irrigation water has steadily gained in popularity since it was first used in the 1930s. Factors contributing to the growth of this system of liquid feeding have been:

(a) the large increase in the cost of labour;
(b) the need to obtain the maximum growth rate and the greatest number of plants per year from each glasshouse;
(c) the development of automatic or semi-automatic systems of pot watering, which, together with fertilizer injectors or diluters, provide an efficient and easy means of combining feeding with watering;
(d) the increased use of loamless mixes, which have a greater need for liquid feeding than the older, loam-based composts.

9.1 PRINCIPLES OF FEEDING

The need for supplementary liquid feeding of container plants largely depends upon:

(a) how long the plant is to be kept in the container;
(b) its vigour and nutrient requirements in relation to the volume of medium or container size;
(c) whether controlled-release fertilizers were used in the mix;
(d) the system of irrigation – with capillary irrigation there is no loss of nutrients whereas with surface irrigation nutrients are easily leached from loamless media.

Most lightweight media have low cation exchange capacities or 'reserve' amounts of nutrients, consequently plants will show deficiency symptoms more rapidly than those grown in loam-based media.

9.1.1 N:K ratios and fertilizer strengths

The preferred balance of N:K in the fertilizer, its concentration and frequency of feeding varies with the crop, the environment and type of management. Some examples of the ratios found in the tissues of crops, and their response to fertilizer concentrations are given below.

Crop		Ratio of N:K in tissue		
		N	K	
Saintpaulia		1	3.02	Hansen (1978)
Peperomia		1	2.67	
Kalanchoe		1	1.12	
Solanum		1	0.49	
Pot Chrysanthemum, base-fertilizer plus liquid feed	summer	1	0.98	Bunt (1976b)
	winter	1	1.06	

Adams & Massey (1984) reported that when tomato plants were grown in recirculating nutrient solution the ratio of absorbed N:K changed with the age of the plants. In the early vegetative stage the ratio was 1 part nitrogen:1.2 parts potassium. Later when the fruit was developing, growth was checked; the uptake of nitrogen decreased and potassium increased, giving an absorption ratio of 1:2.5.

In practice, where a mixed range of container crops is grown, a compromise balance of 1:1 (N:K) is often used, which is varied to meet specific crop and seasonal requirements.

Plant species also differ in their response to the strength of the liquid feed. Boertje (1980) grew six bedding plant species for three months in a mixture of 60% black decomposed peat and 40% sphagnum peat, fertilized with 1.5 kg m^{-3} of a compound fertilizer ('PG mix': 14% N, 16% P_2O_5, 18% K_2O). Liquid feeding commenced after four weeks with the fertilizer 'Kristalon Blue' (17% N, 6% P_2O_5, 18% K_2O) applied at rates of 0.5–4.5 g l^{-1}. The optimal strength, based on visual assessment, was 1.5 g l^{-1} (255 p.p.m. nitrogen, 39 p.p.m. phosphorus, 224 p.p.m. potassium; electrical conductivity (EC) 2.2 mS at 25 °C) for Petunia, Salvia and Ageratum; 3 g l^{-1} for Impatiens, with 3 g l^{-1} and 4.5 g l^{-1} equal best for Dahlia. Tagetes showed leaf scorch at 1 g l^{-1} and 1.5 g l^{-1}, other species did not. Poole & Henley (1981) found that Brassaia, Chamaedorea and Peperomia grew best with a liquid feed having 250 mg l^{-1} nitrogen, 110 mg l^{-1} phosphorus and 200 mg l^{-1} potassium. Increasing the strength to twice and three times this rate reduced their growth rates but did not affect Philodendron and Maranta.

Plants requiring high nutrient levels, e.g. pot chrysanthemums, usually respond to liquid feeding even when controlled-release fertilizers have been used. Suggested strengths of feeds for certain crops are given in Table 9.1.

9.1.2 Interaction with season

Frequently, growers vary the strength of the feed with the season. For example, the first liquid feed given to early tomato plants before they are ready for planting into peat modules will have an EC of about 2000 μS cm^{-1} (2 mmho cm^{-1}). This is gradually increased to 4500 μS cm^{-1} or more at planting time (early January), it is then gradually reduced to 3000 μS cm^{-1} in early March and maintained at 2250–2500 μS cm^{-1} over the remainder of the season. The increase in feed strength is used to reduce the plant vigour and assist in setting the fruit under the poor light conditions of winter. Growers of pot plants will sometimes use high strength liquid feeds to control plant size and leaf colour. These are two examples of 'osmotic growing'; however, this technique should not be practised until growers have some experience of the likely response of the crop.

Table 9.1 Liquid feed strengths for constant feeding.

Species	Nutrient strength (p.p.m. in diluted feed)			Remarks
	nitrogen	phosphorus	potassium	
Azalea	50–100	10–20	20–50	ammonium preferred to nitrate nitrogen; maintain pH at 5.0
bedding plants	200	30	150	
Chrysanthemum				
summer	200	30	150	hand watered or drip system
winter	150	20	200	
Cyclamen	150	20	200	large pots
	60	15	80	mini pots
foliage plants	150–200	20–30	100–150	
Hydrangea	100–200	15–30	100–200	for growing
	50–100	5	200	forcing blue varieties
nursery stock	200	20	150	include 20 p.p.m. Mg
Poinsettia	250	20	150	starting ⎱ plus 0.1 p.p.m.
	150	20	200	finishing ⎰ molybdenum avoid high NH$_4$-N
Rieger Begonia	100–150	15	50–125	
Saintpaulia	75	20	125	nitrate nitrogen preferred; use low nitrogen for variegated forms
tomato	100	15	280	after 5th true leaf expanded
	140	20	375	from planting out
	180	20	375	six weeks after planting
	225	20	375	main season – in peat bags; include 15 p.p.m. Mg

Season also affects the rate of evapotranspiration, its relationship with nutrient uptake by the plant and the leaching requirement. At the Glasshouse Crops Research Institute the daily evapotranspirational loss of water was recorded from chrysanthemums grown with 5 cuttings in 14 cm ($5\frac{1}{2}$ in) one litre pots. Each pot was watered and fed from its own capillary irrigation unit, there was no loss of water from the bench by evaporation or nutrients from the pot by leaching. Over the 11 week period of each crop the evapotranspirational loss ranged from 5357 cm^3 in winter to 17 328 cm^3 in summer. The total nitrogen uptake by the plants was 728 mg in winter and 1262 mg in summer (Bunt, unpublished). Thus the ratio of nitrogen uptake to water loss was 1 mg nitrogen:7.36 cm^3 of evapotranspiration in winter, and 1:13.73 in summer; i.e. the amount of nutrient uptake per unit of evapotranspiration was approximately twice as high in winter as it was in summer. It is therefore probable that the salinity of the media would increase more rapidly in summer than in winter, unless the media is heavily leached. This hypothesis is confirmed by the results of another experiment with differing strengths of liquid feed (Fig. 9.1). The plants had been grown with surface irrigation, but the amount of leaching was obviously insufficient to prevent the salinity from rising more rapidly in summer than in winter.

These results show that in summer or under conditions of high evapotranspiration the nutrient concentration in liquid feeds should be reduced.

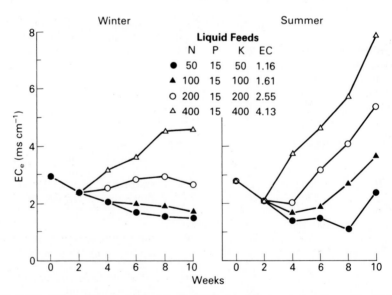

Figure 9.1 Effect of the strength of the liquid feed and the season on the media salinity of pot-grown chrysanthemums (measured as electrical conductivity of a saturated media extract). Note the higher salinity in summer than in winter and the rapid increase in salinity at the end of the crop.

9.1.3 Type of irrigation system

The absence of leaching with capillary irrigation systems means that lower strength liquid feeds will be required. Bunt (1973b) found that, in winter, chrysanthemums irrigated with 100 p.p.m. nitrogen by capillary watering had a similar flowering response to plants having 300 p.p.m. nitrogen by surface watering. As the moisture stress of capillary irrigated media remains relatively constant, short term changes in salt stress will be minimal. However, with surface irrigation the moisture content of the medium can fall substantially between irrigations. This causes a large rise in the salt concentration, consequently the increase in the total soil moisture stress is much greater than the simple reduction in water content would at first suggest. For this reason, media should always be kept moist when growing with high salt concentrations.

9.2   FORMULATING LIQUID FEEDS

Growers of pot plants have the choice of using either proprietary liquid feeds or making up the feeds themselves from soluble chemicals. It will be assumed that in either case the feed will be in a concentrated form and will then be diluted by means of suitable equipment.

9.2.1 Proprietary feeds

One of the principal advantages of using proprietary feeds is convenience. The concentrated feed is already prepared and it is not necessary to weigh out the individual chemicals, in some cases the feeds are supplied in liquid form. Advantage can also be taken of the advanced technology used by the manufacturers in formulating the fertilizers. Some of the materials they use are either not available or are not practical for growers to use. For example, phosphoric acid and ammonia sometimes form the basis of proprietary liquid feeds, with urea being used as a secondary source of nitrogen. Other materials used by commercial manufacturers are ammonium hydroxide, chelated macro- and microelements and growth regulators such as indolbutyric acid. In the USA, superphosphoric acid and ammonium polyphosphates are also used in some proprietary feeds. The advantages of using such feeds are: higher ratios of plant nutrients, resistance to 'salting-out' at low temperatures and the prevention of minor element precipitation by chelation. Some care must be exercised, however, over the use of liquid feeds containing high concentrations of chelated microelements; for example, manganese toxicity could result if the feed contains too much chelated manganese. Present experience suggests that the amount of chelated manganese in the diluted feed should not be more than about 0.5 p.p.m. Feeds containing high levels of microelements should not be used indiscriminately; they should be used only when there is evidence of a deficiency.

In the USA, it is possible to purchase a wide range of liquid feeds formulated specifically for certain crops or for use with special composts. For example, Peters Fertilizers of Pennsylvania, USA, produce, for azalea growers, a liquid feed with a strongly acidifying action. A 'Peat–Lite Special' fertilizer with 15% N, 11% P_2O_5, 29% K_2O is also formulated for use with peat–vermiculite and peat–perlite mixes.

9.2.2 Nursery mixed feeds

An alternative to buying proprietary, ready-mixed feeds is to make up concentrated fertilizer solutions at the nursery. These have the advantages that (a) they are cheaper and (b) it is then easier to follow non-commercial advice obtained from National or State Advisers. Growers sometimes have difficulty translating liquid feed programmes given by advisers into terms of commercial liquid feeds. By dissolving the correct amounts of a few chemicals in water, it is possible for the grower to prepare, easily and cheaply, liquid feeds that meet most of his requirements. Some of the materials which can be used to make the feeds, and their solubilities, are listed in Table 9.2. Liquid feeds suitable for most purposes can be formulated from ammonium nitrate, monoammonium phosphate and potassium nitrate. Urea can be used in place of ammonium nitrate where it is desired to reduce the amount of nitrate nitrogen present. The weights of fertilizers required to make one litre of stock solution which then has to be diluted at 1 in 200, i.e. one part of stock

Table 9.2 Solubility of fertilizers in cold and hot water.

Fertilizer	Solubility g 100 cm^{-3} water		lb gal^{-1} of cold water
	cold	hot	
ammonium nitrate	118.3[0]	871.0[100]	11.7
ammonium sulphate	70.6[0]	103.8[100]	7.0
calcium nitrate	102.5[0]	376.0[100]	10.0
urea	78.0		7.5
monoammonium phosphate	22.7[0]	173.2[100]	2.2
diammonium phosphate	57.5[0]	106.0[70]	5.6
potassium carbonate	112.0[20]	156.0[100]	11.0
potassium chloride	34.7[20]	56.7[100]	3.4
potassium nitrate	13.3[0]	247.0[100]	1.3
potassium sulphate	12.0[25]	24.0[100]	1.2
potassium orthophosphate	90.0[20]		8.9
monopotassium phosphate	167.0[20]		16.5
magnesium sulphate	26.0[0]	73.8[100]	2.5
sodium borate (borax)	1.6[10]	14.2[55]	0.15
Solubor	4.5[10]	32.0[50]	0.44
copper sulphate	31.6[0]	203.3[100]	3.1
manganese sulphate	105.3[0]	111.2[54]	10.4
ferrous sulphate	15.6	48.6[50]	1.5
sodium molybdate	56.2[0]	115.5[100]	5.5

The superscript figures are the temperatures (°C) at which the solubilities were determined.

solution made up to 200 parts with water, to give feeds ranging in strength from 50–300 p.p.m. of nitrogen and potassium are given in Table 9.3. For practical purposes, the phosphorus level of these feeds has been fixed in relation to the nitrogen at a rate of 15 p.p.m. phosphorus to each 100 p.p.m. nitrogen. When potassium nitrate is used as the source of potassium, the relatively high nitrogen content of the fertilizer makes it

Table 9.3 Weight of fertilizers in grams required to prepare one litre of stock solution for dilution at 1 in 200 to give a range of liquid feeds.

Fertilizer	gl^{-1}						
	N 50 P 7.5	N 100 P 15	N 150 P 22.5	N 200 P 30	N 250 P 37.5	N 300 P 45	
ammonium nitrate	16.0	42.3	68.9	95.4	121.8	148.3	K 50
monoammonium phosphate	6.1	12.3	18.4	24.5	30.6	36.7	
potassium nitrate	26.3	26.3	26.3	26.3	26.3	26.3	
(EC)	(0.32)	(0.57)	(0.80)	(1.04)	(1.28)	(1.51)	
ammonium nitrate	5.4	31.9	58.4	84.8	111.3	137.8	K 100
monoammonium phosphate	6.1	12.3	18.4	24.5	30.6	36.7	
potassium nitrate	52.6	52.6	52.6	52.6	52.6	52.6	
(EC)	(0.41)	(0.65)	(0.89)	(1.12)	(1.35)	(1.58)	
ammonium nitrate	*	21.4	47.8	74.3	100.8	127.3	K 150
monoammonium phosphate	6.1	12.3	18.4	24.5	30.6	36.7	
potassium nitrate	78.9	78.9	78.9	78.9	78.9	78.9	
(EC)	(0.58)	(0.73)	(0.96)	(1.20)	(1.43)	(1.66)	
ammonium nitrate		10.8	37.3	63.8	90.2	116.7	K 200
monoammonium phosphate		12.3	18.4	24.5	30.6	36.7	
potassium nitrate		105.3	105.3	105.3	105.3	105.3	
(EC)		(0.82)	(1.04)	(1.29)	(1.52)	(1.74)	
ammonium nitrate		nil	26.8	53.3	79.8	106.2	K 250
monoammonium phosphate		12.3	18.4	24.5	30.6	36.7	
potassium nitrate		131.5	131.5	131.5	131.5	131.5	
(EC)		(0.90)	(1.11)	(1.37)	(1.60)	(1.84)	
ammonium nitrate		†	16.3	42.8	69.2	95.7	K 300
monoammonium phosphate		12.3	18.4	24.5	30.6	36.7	
potassium nitrate		157.8	157.8	157.8	157.8	157.8	
(EC)			(1.22)	(1.44)	(1.69)	(1.87)	

* = 58.9 p.p.m. nitrogen.

† = 117.8 p.p.m. nitrogen.

This table was computed using the following fertilizer analysis: ammonium nitrate 35% N; monoammonium phosphate 12% N, 24.5% P; potassium nitrate 14% N, 38% K. The electrical conductivity (EC) values of the *diluted* fertilizer solutions have been expressed in mmho cm^{-1} at 25 °C.

impossible to prepare feeds having a high potassium and low nitrogen level. If it was intended to make a feed having 300 p.p.m. potassium and 50 p.p.m. nitrogen, it would be necessary to use another source of potassium which does not contain any nitrogen, e.g. potassium chloride or a suitable grade of potassium sulphate.

The weights of fertilizer required to prepare the feeds in Imperial measure, i.e. ounces per gallon, are given in Table 9.4 and in ounces per US gallon in Table 9.5.

9.2.3 Formulae for calculating liquid feeds

Growers, who wish to use liquid feeds of different analyses or composition to those already given in Table 9.3 (or 9.4 & 9.5), can quite easily make up the feeds by using the following formulae. It has been assumed that a concentrated stock solution will first be made, which will then be diluted to the correct strength by means of either a diluter, a mixer-proportioner, an injector or similar device. All these appliances work on a volume basis,

Table 9.4 Weight of fertilizers in ounces to make 1 gal (Imperial) of stock solution for dilution at 1 in 200 to give a range of liquid feeds.

Fertilizer	oz per Imperial gal						
	N 50 P 7.5	N 100 P 15	N 150 P 22.5	N 200 P 30	N 250 P 37.5	N 300 P 45	
ammonium nitrate	2.56	6.78	11.05	15.30	19.53	23.78	⎫
monoammonium phosphate	0.98	1.97	2.95	3.93	4.91	5.89	⎬ K 50
potassium nitrate	4.22	4.22	4.22	4.22	4.22	4.22	⎭
ammonium nitrate	0.87	5.12	9.37	13.60	17.85	22.09	⎫
monoammonium phosphate	0.98	1.97	2.95	3.93	4.91	5.89	⎬ K 100
potassium nitrate	8.43	8.43	8.43	8.43	8.43	8.43	⎭
ammonium nitrate	*	3.43	7.67	11.91	16.16	20.41	⎫
monoammonium phosphate	0.98	1.97	2.95	3.93	4.91	5.89	⎬ K 150
potassium nitrate	12.65	12.65	12.65	12.65	12.65	12.65	⎭
ammonium nitrate		1.73	5.98	10.23	14.46	18.71	⎫
monoammonium phosphate		1.97	2.95	3.93	4.91	5.89	⎬ K 200
potassium nitrate		16.89	16.89	16.89	16.89	16.89	⎭
ammonium nitrate		nil	4.30	8.55	12.80	17.03	⎫
monoammonium phosphate		1.97	2.95	3.93	4.91	5.89	⎬ K 250
potassium nitrate		21.09	21.09	21.09	21.09	21.09	⎭
ammonium nitrate		†	2.61	6.86	11.10	15.35	⎫
monoammonium phosphate		1.97	2.95	3.93	4.91	5.89	⎬ K 300
potassium nitrate		25.30	25.30	25.30	25.30	25.30	⎭

* = 58.9 p.p.m. nitrogen.

† = 117.8 p.p.m. nitrogen.

Table 9.5 Weight of fertilizers in ounces to make 1 US gallon of stock solution for dilution at 1 in 200 to give a range of liquid feeds.

Fertilizer	N 50 P 7.5	N 100 P 15	N 150 P 22.5	N 200 P 30	N 250 P 37.5	N 300 P 45	
ammonium nitrate	2.14	5.65	9.20	12.74	16.26	19.80	} K 50
monoammonium phosphate	0.81	1.64	2.46	3.27	4.09	4.90	
potassium nitrate	3.51	3.51	3.51	3.51	3.51	3.51	
ammonium nitrate	0.72	4.26	7.80	11.32	14.86	18.40	} K 100
monoammonium phosphate	0.81	1.64	2.46	3.27	4.09	4.90	
potassium nitrate	7.02	7.02	7.02	7.02	7.02	7.02	
ammonium nitrate	*	2.86	6.38	9.92	13.46	17.00	} K 150
monoammonium phosphate	0.81	1.64	2.46	3.27	4.09	4.90	
potassium nitrate	10.54	10.54	10.54	10.54	10.54	10.54	
ammonium nitrate		1.44	4.98	8.52	12.04	15.58	} K 200
monoammonium phosphate		1.64	2.46	3.27	4.09	4.90	
potassium nitrate		14.06	14.06	14.06	14.06	14.06	
ammonium nitrate		nil	3.58	7.12	10.65	14.18	} K 250
monoammonium phosphate		1.64	2.46	3.27	4.09	4.90	
potassium nitrate		17.56	17.56	17.56	17.56	17.56	
ammonium nitrate		†	2.18	5.72	9.24	12.78	} K 300
monoammonium phosphate		1.64	2.46	3.27	4.09	4.90	
potassium nitrate		21.07	21.07	21.07	21.07	21.07	

oz per US gal

* = 58.9 p.p.m. nitrogen.
† = 117.8 p.p.m. nitrogen.

which is much more practical than the weight basis. The density or specific gravity of the stock solution will vary with the fertilizers used and the concentration; the density can be 15% or more greater than that of plain water. All calculations have, therefore, been made on the basis of a *weight* of fertilizer being applied to a *volume* of water, i.e. w/v.

The metric system of calculation is much simpler to use than Imperial measures, once the grower is familiar with the units, but to assist those growers who may be unable to make the change to metric with their present equipment, worked examples have also been given in Imperial measures.

To convert to USA measures, substitute 75.0 for the factor 62.5 in the calculations using Imperial measures.

Calculation of weight of fertilizer required to provide a given concentration (p.p.m.) in the diluted feed To calculate the weight of ammonium sulphate (21% nitrogen) required to make a stock solution which, after being diluted at 1 in 200, gives a liquid feed of 100 p.p.m. nitrogen:

Metric

$$\begin{matrix}\text{grams of} \\ \text{fertilizer} \\ \text{per litre of} \\ \text{stock solution}\end{matrix} = \frac{\text{p.p.m. of nutrient} \times \text{dilution rate}}{1000} \times \frac{100}{\begin{matrix}\text{\% nutrient in} \\ \text{fertilizer}\end{matrix}}$$

$$= \frac{100 \times 200}{1000} \times \frac{100}{21} = 95.24 \text{ g}$$

Imperial measure

$$\begin{matrix}\text{ounces of fertilizer per} \\ \text{gallon of stock solution}\end{matrix} = \frac{\text{p.p.m. of nutrient required} \times \text{dilution}}{62.5 \times \text{\% nutrient in fertilizer}}$$

$$= \frac{100 \times 200}{62.5 \times 21} = 15.24 \text{ oz}$$

To calculate the concentration (p.p.m.) of nutrient in the dilute feed supplied by a given amount of fertilizer If 25 g of urea (46% nitrogen) is dissolved in 1 litre of water and diluted 1 in 200, what p.p.m. of nitrogen does this supply?

Metric

$$\begin{matrix}\text{p.p.m. in} \\ \text{dilute feed}\end{matrix} = \frac{\text{fertilizer weight} \times 1000}{\text{dilution}} \times \frac{\text{\% nutrient}}{100}$$

$$= \frac{25 \times 1000}{200} \times \frac{46}{100} = 57.5 \text{ p.p.m. nitrogen}$$

Imperial

e.g. 16 oz of urea dissolved in 1 gal and diluted at 1 in 200.

$$\begin{matrix}\text{p.p.m. in} \\ \text{dilute feed}\end{matrix} = \frac{\text{oz fertilizer per gal} \times \text{\% nutrient} \times 62.5}{\text{dilution}}$$

$$= \frac{16 \times 46 \times 62.5}{200} = 230 \text{ p.p.m. nitrogen}$$

To calculate the amounts of fertilizers required to make up a complete NPK feed where fertilizers supplying more than one nutrient are used How to make up 1 litre of stock solution to be diluted at 1 in 200 to give 200 p.p.m. nitrogen, 30 p.p.m. phosphorus and 150 p.p.m. potassium in the dilute feed using:

> monoammonium phosphate (12% nitrogen, 24.5% phosphorus)
> potassium nitrate   (14% nitrogen, 38% potassium)
> ammonium nitrate   (35% nitrogen)

Using the formulae already given above,

(1) calculate the amount of monoammonium phosphate required to supply 30 p.p.m. phosphorus;

(2) calculate the amount of nitrogen this also supplies;
(3) calculate the amount of potassium nitrate required to supply 150 p.p.m. potassium;
(4) calculate the amount of nitrogen this also supplies;
(5) add the amounts of nitrogen supplied by the monoammonium phosphate and the potassium nitrate and subtract this from the required value of 200 p.p.m. nitrogen;
(6) calculate the amount of ammonium nitrate required to supply this amount of nitrogen.

Metric

(1) weight of monoammonium phosphate required $= \dfrac{30 \times 200}{1000} \times \dfrac{100}{24.5} = 24.5$ g

(2) this also supplies nitrogen $= \dfrac{24.5 \times 1000}{200} \times \dfrac{12}{100} = 14.7$ p.p.m.

(3) weight of potassium nitrate required $= \dfrac{150 \times 200}{1000} \times \dfrac{100}{38} = 78.9$ g.

(4) this also supplies nitrogen $= \dfrac{78.9 \times 1000}{200} \times \dfrac{14}{100} = 55.2$ p.p.m. nitrogen

(5) total nitrogen so far supplied $= 14.7 + 55.2 = 69.9$ p.p.m. nitrogen

$200 - 69.9 = 130.1$ p.p.m. nitrogen still required

(6) weight of ammonium nitrate required $= \dfrac{130.1 \times 200}{1000} \times \dfrac{100}{35} = 74.3$ g

Therefore to make this liquid feed requires:

24.5 g monoammonium phosphate
78.9 g potassium nitrate
74.3 g ammonium nitrate

Imperial

(1) weight of monoammonium phosphate required $= \dfrac{30 \times 200}{62.5^* \times 24.5} = 3.9$ oz

(2) this also supplies nitrogen $= \dfrac{3.9 \times 12 \times 62.5^*}{200} = 14.7$ p.p.m. nitrogen

(3) weight of potassium nitrate required $= \dfrac{150 \times 200}{62.5^* \times 38} = 12.6$ oz

(4) this also supplies
     nitrogen $= \dfrac{12.6 \times 14 \times 62.5^*}{200} = 55.2$ p.p.m.
     nitrogen

(5) total nitrogen
     supplied so far $= 14.7 + 55.2 = 69.9$ p.p.m. nitrogen

$200 - 69.9 = 130.1$ p.p.m. nitrogen still required

(6) weight of ammonium
     nitrate $= \dfrac{130.1 \times 200}{62.5^* \times 35} = 11.9$ oz

In Imperial measure, this feed requires:

> 3.9 oz monoammonium phosphate
> 12.6 oz potassium nitrate
> 11.9 oz ammonium nitrate

in 1 gal of water, dilute at 1 in 200.

* (For USA substitute 75.0 for 62.5).

Proprietary fertilizers Calculation of the nutrients present when a proprietary fertilizer is diluted. For example, a proprietary fertilizer in concentrated form contains 6% nitrogen and is diluted at 1 in 200; what will be the amount of nitrogen in the diluted feed?

$$\text{p.p.m. in feed} = \frac{\% \text{ nutrient (w/v) in concentrate} \times 10\,000}{\text{dilution}}$$

$$= \frac{6 \times 10\,000}{200} = 300 \text{ p.p.m. nitrogen}$$

Calculation of the dilution necessary to give the required nutrient concentration. For example, the concentrated fertilizer contains 8% (w/v) nitrogen; what dilution is required to give 200 p.p.m.?

$$\text{dilution required} = \frac{\% \text{ nutrient (w/v)} \times 10\,000}{\text{p.p.m. required}}$$

$$= \frac{8 \times 10\,000}{200} = 400$$

i.e. dilute at 1 in 400.

9.2.4 USA measures and practice

The US gallon is approximately equal to 0.83 or $\frac{5}{6}$ of an Imperial gallon and the amount of fertilizers used in making up liquid feeds must be adjusted accordingly; more precise conversions of US to metric and US to Imperial liquid measures are given in Tables 1 and 2 in the Appendices (pp. 284–285). The necessary alterations in the weights of fertilizers required to make up the complete range of stock solutions based on ammonium nitrate, monoammonium phosphate and potassium nitrate are given in Table 9.5. The salinity of the feeds will, of course, be the same as for the

Table 9.6 Preparation of liquid feeds from complete NPK water-soluble fertilizers (based on US gal).

Fertilizer			Element equivalent (%)			Ounces fertilizer per US gallon diluted 1 in 200	p.p.m. of		
N	P_2O_5	K_2O	N	P	K		N	P	K
20	20	20	20	8.8	16.6	26.6	200	88	166
15	30	15	15	13.2	12.5	35.5	200	176	167
14	14	14	14	6.2	11.6	38.0	200	88	166
21	7	7	21	3.1	5.8	25.3	200	30	55
20	5	30	20	2.2	24.9	26.6	200	22	249
25	10	10	25	4.4	8.3	21.3	200	35	66

To obtain 100 p.p.m. of nitrogen use half the above weights of fertilizer, and for 150 p.p.m. nitrogen use three-quarters of the weight.

corresponding feeds in Table 9.3. Some models of Solu-Bridge conductivity meters measure the EC in units of 10^{-3} whereas other models are calibrated in units of 10^{-5}. To convert from one unit to another move the decimal point two places, i.e. an EC reading of 0.75×10^{-3} measured on a recent model is equivalent to 75×10^{-5} measured on an older model.

It is also possible for growers in the USA to make up concentrated stock solutions from water-soluble fertilizers having a range of NPK ratios, and examples of their use are given in Table 9.6. In the US the chemical composition of irrigation water, and sometimes of liquid feeds, is reported as $meq\,l^{-1}$ (milliequivalents per litre). Examples of the nutrient concentrations of fertilizer solutions in p.p.m. or $meq\,l^{-1}$ are given in Tables 9.7 and 9.8. For *dilute* fertilizer solutions, concentrations expressed in p.p.m. and $mg\,l^{-1}$ are similar.

Table 9.7 Nutrient concentrations in p.p.m. and $meq\,l^{-1}$ supplied by fertilizers at either 100 g per 1000 l or 1 lb per 1000 US gal.

Fertilizer	Analysis (%)	100 g per 1000 l*		1 lb per 100 US gal	
		p.p.m.	$mEq\,l^{-1}$	p.p.m.	$mEq\,l^{-1}$
Ammonium nitrate (NH_4NO_3)	35 N	17.5 NH_4-N	1.25 NH_4^+-N	21 NH_4-N	1.50 NH_4^+-N
		17.5 NO_3-N	1.25 NO_3^--N	21 NO_3-N	1.50 NO_3^--N
Monoammonium phosphate ($NH_4H_2PO_4$)	12 N	12 N	0.86 NH_4^+-N	14 N	1.0 NH_4^+-N
	25 P	25 P	0.86 $H_2PO_4^-$-P	30 P	1.0 $H_2PO_4^-$-P
Potassium nitrate (KNO_3)	13 N	13 N	0.92 NO_3^--N	15 N	1.10 NO_3^--N
	38 K	38 K	0.92 K^+	45 K	1.10 K^+
Calcium nitrate ($Ca(NO_3)_2 \cdot 4H_2O$)	12 N	12 N	0.85 NO_3^--N	14 N	1.00 NO_4^--N
	17 Ca	17 Ca	0.85 Ca^{2+}	20 Ca	1.00 Ca^{2+}

* 1 lb in 1000 Imperial gallons gives the same concentration as 100 g in 1000 litres.

Table 9.8 Comparison of nutrient concentrations in p.p.m. (or mg l^{-1}) and mEq l^{-1}.

p.p.m. (mg l^{-1})	mEq l^{-1}	Equivalent weight
nitrogen 100 (as ammonium nitrogen)	7.14 NH_4^+-N	14.0
nitrogen 100 (as nitrate nitrogen)	7.14 NO_3^--N	14.0
phosphorus 100	3.22 $H_2PO_4^-$-P	31.0
potassium 100	2.55 K^+	39.1
calcium 100	4.98 Ca^{2+}	20.04
bicarbonate 100	1.64 HCO_3^-	61.0
magnesium 100	8.23 Mg^{2+}	12.16

To convert mEq l^{-1} to p.p.m. multiply the number of milliequivalents of the ion by its equivalent weight.

To convert p.p.m. to milliequivalents divide by the equivalent weight.

For *dilute* nutrient solutions, where the specific gravity of the solution is virtually 1.0, p.p.m. = mg l^{-1}.

In Florida, rates for liquid feeding container crops are sometimes given as pounds of fertilizer per acre or per 1000 ft^2. For example, the suggested rate of fertilizer for *Brassaia actinophylla* is 18 lb water-soluble 20–20–20 per 1000 ft^2 per month. The relationship between the fertilizer concentration and volume of irrigation solution is shown in Table 9.9. If 0.5 inches of 20% nitrogen fertilizer solution at 100 p.p.m. is applied to 1000 ft^2, this equals 1.31 lb of fertilizer. The volume of fertilizer solution received by pots of different size is given in Table 9.10.

Calcium nitrate is used in nursery mixed feeds in the USA to a much greater extent than in Britain. The calcium in the fertilizer prevents the pH of the mix from falling too low and the nitrogen is all in the nitrate form. A feed having all the nitrogen in the nitrate form can be made from:

	oz per US gal
potassium nitrate	14.75
calcium nitrate	21.00

Dilute at 1 in 200 to give 200 p.p.m. nitrogen and 200 p.p.m. potassium.

Table 9.9 The applied weight of fertilizer (lb of 20% nitrogen fertilizer per 1000 ft^2) resulting from different combinations of the strength of feed and the amount of irrigation (Biamonte 1977).

Inches of fertilizer solution	Concentration of nitrogen (p.p.m.)				
	50	75	100	150	200
0.25	0.32	0.49	0.65	0.97	1.29
0.5	0.64	0.98	1.31	1.95	2.59
1.0	1.28	1.95	2.61	3.89	5.19

Table 9.10 Equivalent rates of irrigation over a range of pot sizes.

Amount of solution applied (in)	Equivalent rate (gal per 1000 ft^2)	Amount of solution per pot (fl oz)				
		4" pot	6" pot	8" pot	10" pot	12" pot
0.25	155.8	1.7	4.0	7.0	11.0	15.7
0.5	311.7	3.5	7.8	14.0	21.8	31.3
1.0	623.4	7.0	15.7	28.0	43.5	62.7

In Britain the calcium present in most water supplies usually results in the pH of the mix rising rather than falling (see Table 9.11). Guidelines for using calcium nitrate in mixed feeds are given in section 9.3.1. Some samples of calcium nitrate produce a scum when dissolved to make up the concentrate. This is caused by the paraffin wax and oil used as conditioning agents to reduce water vapour absorption during the manufacture of calcium nitrate. A wetting agent can be used to emulsify the paraffin and allow it to pass through the injector.

9.2.5 Liquid feed strength and salinity

Liquid fertilizers are dilute salt solutions and therefore have a salinity value. For each liquid feed, irrespective of whether it is based on a single salt or a mixture of salts, there is a direct relationship between the strength of the solution and its salinity value. The salinity of a liquid feed is most easily determined from its electrical conductivity (EC) by means of a small conductivity meter. Once this value is known, it can be used to monitor the strength of the feed being applied. The EC values of the feeds given in Table 9.3 have been included in parentheses; these values are for the *diluted* feeds. It must be remembered that the conductivity value of the plain water used to dilute the concentrated feed must be added to the values in the table. For example, the 200 p.p.m. nitrogen, 30 p.p.m. phosphorus, 150 p.p.m. potassium feed has an EC value of 1.20 mS (mmho). The public water supply at the Glasshouse Crops Research Institute has an EC of 0.48 mS, and water from the Institute's borehole has an EC of 0.75 mS. Consequently, the EC value of this particular feed will be 1.68 mS if the public water supply is used and 1.95 mS if the borehole water is used.

For practical purposes, the EC value of a liquid feed can be calculated from the separate values of the various fertilizers, once these are known. This relationship has been determined for a number of fertilizers commonly used to prepare liquid feeds and is given in Figure 9.2. An example of using this method to compute the EC value of a liquid feed is as follows: What is the EC of a feed, having 200 p.p.m. nitrogen, 30 p.p.m. phosphorus and 150 p.p.m. potassium, made from ammonium nitrate, monoammonium phosphate and potassium nitrate?

Fertilizer solution	EC value (mmho or mS cm^{-1})
(1) the potassium nitrate at 150 p.p.m. potassium (this also supplies 55 p.p.m. nitrogen)	0.55
(2) the monoammonium phosphate at 30 p.p.m. phosphorus (this also supplies 15 p.p.m. nitrogen)	0.11
(3) the remaining 130 p.p.m. nitrogen (supplied by the ammonium nitrate)	0.62
Total EC	1.28

To this must be added the EC of the water supply.

If desired, the ratio of ammonium to nitrate in the above feed could be changed by using 56.6 g of urea (46% nitrogen) in place of the 74.3 g of ammonium nitrate (35% nitrogen). The feed will then contain 72% of ammonium nitrogen and 28% nitrate nitrogen in place of the previous ratio of 40% ammonium nitrogen and 60% nitrate nitrogen. The EC of the diluted feed would then be 0.73 mS cm^{-1}, because the urea in solution does not ionize (Fig. 9.2).

Advisory or Extension workers in Holland suggest that an approximate estimate of the EC of the dilute feed (in μS) can be obtained if the

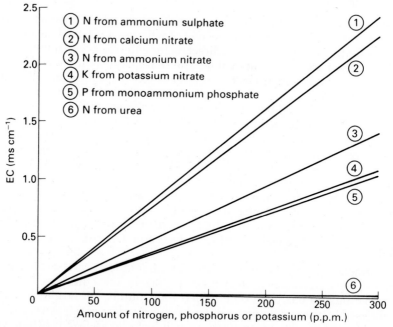

(1) N from ammonium sulphate
(2) N from calcium nitrate
(3) N from ammonium nitrate
(4) K from potassium nitrate
(5) P from monoammonium phosphate
(6) N from urea

Figure 9.2 Relationship of salinity of the liquid feed to the amount of nitrogen, phosphorus or potassium supplied by different fertilizers. Nitrogen supplied by ammonium sulphate, calcium nitrate, ammonium nitrate, urea. Phosphorus supplied by monoammonium phosphate. Potassium supplied by potassium nitrate.

216

concentrations (in meq l^{-1}) of potassium, calcium, magnesium, sodium and ammonium are summed and multiplied by 100. However, it is advisable to monitor the EC as already described.

9.2.6 Phosphorus in liquid feeds

When pot plants were grown in composts based on mineral soils, phosphorus was frequently omitted from the liquid feed. This was because most mineral soils are able to retain a large amount of the added phosphorus and also have reserve forms of phosphorus present. In most cases, the addition of superphosphate in the base fertilizer was sufficient to meet the plant's needs. The much greater water-solubility of phosphorus in loamless mixes and the increased risk of its loss by leaching has been examined in Chapter 4. It is desirable, therefore, when using loamless mixes, to include some phosphorus in the liquid feed.

In the UK, horticulturists have been reluctant to include phosphorus in liquid feeds made with alkaline water for fear of causing blockages in trickle or drip irrigation systems. Ways of avoiding or clearing blockages, which can be caused in various ways, are discussed in section 10.1.1. It should be appreciated that most of the calcium compounds found in blocked nozzles will have arisen from evaporation of the water and not from an immediate chemical precipitation caused by the phosphorus. Growers in the USA have used phosphoric acid to acidify their irrigation water, and as a way of supplying phosphorus, for more than two decades (Matkin & Petersen 1971).

The effects of water quality, purity of phosphorus supply, sources of nitrogen and potassium and the inclusion of magnesium in the liquid feed, on the formation of insoluble phosphates were investigated by Bunt (1976a). Water quality only began to affect the formation of insoluble phosphates five days after the concentrated stock solution had been diluted (Fig. 9.3). Some growers are also reluctant to include magnesium in the liquid feed for fear of increasing precipitation. Sulphates of ammonium, potassium and magnesium actually reduced the rates of insoluble phosphorus formation (Fig. 9.4). Phosphoric acid in an 'Analar', or food grade, quality is the preferred source of phosphorus for liquid feeds; if monoammonium phosphate is used it should be made from the 'thermal grade' and not the 'wet process' phosphoric acid. A typical analysis of high grade monoammonium phosphate is:

12.1% N, 26.6% P, 0.03% Fe_2O_3, 0.03% Al_2O_3, 0.04% CaO, 0.02% MgO, 0.2% F, 0.3% SO_3, 0.0% Si, 0.0003% As

Grades of lower purity may produce small amounts of precipitate, this can be prevented by the addition of the disodium salt of EDTA at 15 p.p.m. to the dilute feed. Polyammonium phosphates have not been satisfactory sources of phosphorus in liquid feeds. They can produce white gelatinous precipitates of iron and aluminium phosphates within two hours of dilution.

217

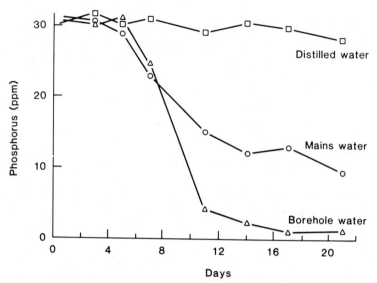

Figure 9.3 Rates of phosphorus precipitation from liquid feeds made with water of differing qualities. Mains water: pH 7.3, total hardness (CaCO$_3$) 259, alkalinity (CaCO$_3$) 208, total solids 362, free carbon dioxide 17 mg l^{-1}. Borehole water: pH 7.4, total hardness 356, alkalinity 250, total solids 528, free carbon dioxide 37 mg l^{-1}.

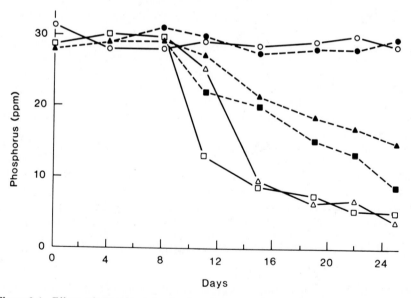

Figure 9.4 Effects of nitrogen and potassium sources and magnesium on the precipitation of phosphorus from liquid feeds.
Key: ○ ammonium sulphate + potassium sulphate; ● ammonium sulphate + potassium sulphate + magnesium sulphate; △ urea + potassium nitrate; ▲ urea + potassium nitrate + magnesium sulphate; □ ammonium nitrate + potassium nitrate; ■ ammonium nitrate + potassium nitrate + magnesium sulphate.

9.2.7   Magnesium in liquid feeds

Magnesium can be included in the liquid feed for such plants as tomato and *Solanum capsicastrum* at the rate of 15 p.p.m. in the dilute feed. This can be obtained by adding magnesium sulphate (Epsom salts, 9.9% magnesium) to the stock solutions at the following rates:

$$\left.\begin{array}{l} 4.9 \text{ oz per Imperial gal} \\ 4.0 \text{ oz per US gal} \\ 30.3 \text{ g l}^{-1} \end{array}\right\} \begin{array}{l} \text{dilute at} \\ 1 \text{ in } 200 \end{array}$$

9.3   PRACTICAL ASPECTS OF FEEDING

9.3.1   Dissolving fertilizers

When preparing stock solutions, it is convenient to use either hot water to dissolve the fertilizers, or to heat the cold water with a steam hose after the fertilizers have been added. Potassium nitrate causes a significant drop in the temperature of the water as it dissolves, it also has a relatively low solubility in cold water (Table 9.2). Sodium borate also has a low solubility, requiring hot water and frequent stirring. It is desirable to add the fertilizers to the water slowly, stirring while they dissolve.

Monoammonium phosphate can be used to make concentrated stock solutions without any risk of precipitates forming as the amount of calcium in the water will be relatively small and the pH of the concentrate will be about 4.5. Calcium nitrate should not, however, be used in concentrates made from fertilizers containing either sulphates or phosphates. Neither should chelates of iron, copper or manganese be added to solutions containing nitric or phosphoric acids.

Where two stock solutions are being injected separately into the irrigation water, monoammonium phosphate, magnesium sulphate and microelements, if required, should be mixed in one tank, with ammonium nitrate, potassium nitrate and calcium nitrate in a second tank. If very low ammonium and high potassium are required, as for early tomato crops, the monoammonium phosphate can be replaced with monopotassium phosphate and the ammonium nitrate replaced with calcium nitrate. However, for many crops some ammonium in the feed helps to increase the availability of the iron and improves leaf colour.

9.3.2   Fertilizer dyes

It is essential, when using displacement type injectors, that the stock solution be coloured to distinguish it from the plain water which is entering the injector. Dyeing the stock solution enables the user to see whether the stock solution has been used up and also whether any mixing of the two liquids is occurring within the container. At certain times, it is also useful to be able to see from the weak colour of the liquid being discharged from the end of the irrigation line that the injector is supplying liquid feed and not plain water. *The colour of the solution cannot, however, be used to check the accuracy of the injector, this can only be done by using a conductivity meter (§ 9.3.3).* Dyes suitable for colouring the stock solution include Disulphine Blue VN 150 (lissamine turquoise VN 150), fluorescein sodium (uranin) and naphthalene orange G 125 (lissamine orange G 125). They are used at the rate of 1 oz to 20 gal of stock solution (1 g per 3 l). It is useful to reserve a separate colour dye for each feed; this provides a ready means of identifying the feed being used, and if white plastic tanks are used to store the stock solutions the amount of feed remaining in the tanks can be seen readily without removing the lids. When using positive displacement type injectors, it is not essential to dye the stock solution.

9.3.3   Control of liquid feed strength

The strength at which the liquid feeds are applied to the plants can be varied in two ways. One method, as shown in Table 9.3, is to make up stock solutions of different strengths and apply them at a fixed dilution rate of 1 in 200. An alternative method is to make up a stock solution of fixed strength and to vary the rate at which it is diluted or injected. Using a dilution rate of 1 in 100 would double the strength of the feed, and a rate of 1 in 400 would halve the strength. Although this method appeals because of its simplicity, it is subject to difficulties. If some stock solutions are increased in strength to allow for a higher dilution rate, there may be trouble over 'salting-out' at low temperatures. Equally, if very weak solutions are made with the intention of using low rates of dilution, say 1 in 50 or 1 in 100, the relatively low specific gravity of the stock solution will mean that there is the risk of its mixing with the incoming displacement water unless the two solutions are separated in the container. If it is desired to use these low rates of dilution, either the two liquids must be physically separated (§ 9.4.1) or a positive-type injector should be used. Also, with some injectors it is either difficult or impossible to change the dilution rate. For these reasons the constant dilution rate at 1 in 200 is preferred.

It has already been seen that each liquid feed has its own salinity or EC value. Once this is known, a small inexpensive conductivity meter can be used to check on the accuracy of the diluting equipment. The determination is easily made by placing a sample of the *diluted* liquid feed in the cell of a direct reading conductivity meter. With some meters, it is necessary to set the meter to the temperature of the liquid, other models have an automatic temperature compensation. It must be remembered that the EC of the plain water must be *added* to the EC values of the feeds given in Table 9.3 and Figure 9.2. If feeds are made up with chemicals other than those used in Table 9.3, all that is required to determine the correct EC of the diluted feed is to accurately make a dilution at the same rate as the injector is set to operate. Flow-type electrodes permanently fitted into the pipe leading from the injector enable quick checks to be made while the equipment is in operation.

9.3.4 Corrosive properties of stock solutions

Although concentrated solutions of some fertilizers can be corrosive, those formulations made up by the grower from the materials at his disposal do not usually cause any serious trouble, and the proprietary liquid fertilizers are manufactured under close control. Hatfield *et al.* (1958) examined the corrosive effect of liquid fertilizers on various metals and concluded that both stainless steel and mild steel were satisfactorily resistant to corrosion. Although some aluminium alloys were found to be satisfactory, others were not. Solutions with a $N:P_2O_5$ ratio of 1:3 were three times as corrosive as solutions with a 1:1 ratio. Potassium was found to give a marked reduction in the amount of corrosion, even when potassium chloride was used. In many instances, the addition of 0.1% of sodium dichromate to the solutions decreased corrosion by 90%. Ammonium thiocyanate is also used as a corrosion inhibitor.

9.3.5 pH control with liquid feeds

The rise in pH of mixes caused by bicarbonates in the irrigation water can be controlled by either acid treatment of the water (§ 9.5.3) or using acidifying fertilizers. An example of the effects which liquid fertilizer composition can have on media pH is shown in Table 9.11. Pot

Table 9.11 Control of media pH of pot chrysanthemums by liquid fertilizer formulation.

Treatment		Media pH, measured in:	
		0.01 M $CaCl_2$	water
pH at start of experiment		5.58	6.15
pH at finish:			
plain water		6.78	7.35
ammonium nitrate	with ammonium sulphate		
monoammonium phosphate	added at:		
potassium nitrate	0.0 p.p.m. nitrogen	6.63	6.83
	32.5 p.p.m. nitrogen	6.25	6.42
	65.0 p.p.m. nitrogen	5.59	5.89
	97.5 p.p.m. nitrogen	4.79	5.20
	130.0 p.p.m. nitrogen	4.08	4.38
ammonium nitrate			
monoammonium phosphate		5.81	6.05
potassium sulphate			
calcium nitrate			
monoammonium phosphate		6.98	7.31
potassium nitrate			

pH determinations were made on a summer crop of chrysanthemums, ten weeks after potting.

All feeds contained 200 p.p.m. nitrogen, 30 p.p.m. phosphorus and 150 p.p.m. potassium, and were made with borehole water containing 250 p.p.m. equivalent calcium carbonate.

In winter, pH changes due to water quality and fertilizer composition are less marked.

chrysanthemums grown in 14 cm pots received the same NPK concentrations from different nitrogen and potassium fertilizers. For situations having similar water quality and media buffer capacity to this experiment, the pH could be controlled by either using ammonium sulphate to supply between 25 and 50 p.p.m. of nitrogen, or using a liquid feed grade of potassium sulphate. In another experiment, the pH was controlled by adding either phosphoric or nitric acid to reduce the pH of the water to 6.10 or 5.75. However, at the finish of the crop mixes having the low and high phosphoric acid treatments had 170 and 245 mg l^{-1} of phosphorus respectively, compared to 29 mg l^{-1} for control treatments. For this reason, nitric acid treatment is preferred to phosphoric acid for long-term crops. Ferrous sulphate at 3 g l^{-1} every seven or 14 days will also control the pH rise, but to avoid deposits on the leaves it must only be applied to the surface of the medium.

9.4 INJECTION EQUIPMENT

Dilute fertilizer solutions suitable for applying to pot plants can be obtained by two means. Either the required weights of fertilizers are dissolved in a measured volume of water and the solution is then ready for applying directly to the plants, or a highly concentrated stock solution is first prepared and this is diluted before use. The former method has the advantage of a high degree of accuracy and the avoidance of problems with the precipitation of certain elements which can occur in concentrated solutions. The principal disadvantages of this system is the need for large storage tanks, which must be lined if the water is to be acidified. The use of fertilizer injectors enables dilute fertilizer solutions to be automatically and accurately made from small volumes of concentrated stock solution.

9.4.1 Differential pressure injectors

Injectors in this group utilize the drop in water pressure when it passes through an orifice, the pressure on the upstream side of an orifice being greater than on the downstream side. This reduced pressure is used to lift some of the concentrated stock fertilizer solution into the mainstream of water and thereby create a dilute nutrient feed. If the stock solution is in an open container the injector is known as a *venturi*, e.g. the 'Hozon'. Large spraying machines also use this principle to fill their tanks from a pond or stream. The rate of injection is affected by the pressure on the upstream side of the injector and also by the distance that the stock solution has to be lifted.

An improved form of injector which also uses the differential pressure principle is the *displacement* injector. The concentrated nutrient solution is placed in a pressure tank, preferably in a rubber or plastic bag within the tank. As some of the high pressure water from the upstream side of the injector is fed between the bag and the tank wall, concentrated solution is drawn into the low pressure side of the injector. Placing the concentrated solution within a bag, e.g. 'Gewa' proportioner, prevents the concentrate and the displacing water from mixing and thereby causing errors in the dilution. The 'Cameron' dilutor relies on the specific gravity of the concentrated stock solution, usually about 1.12–1.15, to prevent rapid diffusion and mixing between it and the displacing water.

9.4.2 Positive injectors

These operate by injecting or pumping the concentrated feed into the water line. The *proportional* injectors are controlled by the rate of water flow, using either water motors, e.g. 'Merit Commander' and 'Smith Measuremix', or a water meter, e.g. 'Baggaley'. These systems take account of varying rates of water flow and give accurate and constant dilution rates. *Metering*, or *dosing*, pumps give a constant rate of injection but do not take account of variations in the water flow rate; they are not recommended for liquid feeding. It is also desirable that the rate of injection should be adjustable rather than being fixed at one rate.

9.4.3 Use of injectors

Most local authorities stipulate that injectors must not be coupled directly into the public water supply. A break must be made in the supply line to guard against the possibility of chemicals siphoning back into the public water supply should there be a reduction in the mains water pressure. Often the installation of check valves in the water main is not regarded by the authority as giving an adequate safeguard against siphoning, and a water storage tank fitted with a ball float control will be required. In the USA, backflow or siphoning is prevented by fitting a vacuum breaker, this is a combination check valve and air relief valve.

   In the interest of economy, growers sometimes use a single large central injector to supply liquid feed to the whole nursery. This has the disadvantage of not being sufficiently flexible to permit different feeds to be used for different plants. In some cases, it is impossible even to obtain a plain water supply in the glasshouse, only liquid feed is available. As already shown, excessive feeding can cause high salinities and thereby restrict plant growth; the ability to apply plain water to plants if required is highly desirable. A water loop to bypass the injector should always be fitted into the system. Not only does this enable plain water to be used if required, in the event of failure the injector can easily be removed from the system for repair.

9.5   QUALITY OF IRRIGATION WATER

9.5.1   Plant response to water quality

Sonneveld & van Beusekom (1974) used increasing concentrations of sodium chloride and a mixture of salts, i.e. calcium chloride, magnesium chloride, magnesium sulphate, sodium sulphate and sodium bicarbonate in such proportions as to resemble the water in the West Netherlands, to study plant response. In the range 0.9–4.5 mmho cm^{-1} the yields of lettuce, tomatoes and cucumber showed a linear decrease with increasing salt concentrations; an increase of 1 mmho cm^{-1} in the salinity reduced the yields by 4, 7 and 14% respectively. However, the yields of spinach were not affected. With flower crops, Sonneveld & Voogt (1983) found that carnation and *Chrysanthemum* were the least sensitive, *Gerbera* and *Hippeastrum* showed medium sensitivity and *Anthurium* had the greatest sensitivity to salinity; the latter also showed a specific sensitivity to sodium chloride. Lunt *et al.* (1956) found that *Azalea* was very sensitive to bicarbonate, which increased the absorption of sodium and decreased the absorption of calcium and iron by this species. *Saintpaulia* is sensitive to quite low amounts of ammonium (Kohl *et al.* 1955) and *Begonia Rex* is particularly sensitive to chloride (Pearson 1949). Wall & Cross (1943) concluded that water containing 200 p.p.m. bicarbonate (3.3 meq l^{-1}) will seldom cause toxicity and 500 p.p.m. (8.2 meq l^{-1}) is the maximum level that can be used without treatment of the water.

The quality of the water can also affect the appearance of the plants, iron and bicarbonate causing leaf spotting; water for overhead watering should not contain >0.4 p.p.m. iron and 210 p.p.m. of carbonate hardness. Leaf residues of these elements have been successfully removed from the foliage of ornamental plants by spraying with 3% oxalic acid plus a surfactant (Broschat & Fitzpatrick 1980). Plants should be rinsed before the acid crystallizes or dries on the foliage.

The importance of good quality irrigation water for container crops will be apparent from the large amounts lost by evapotranspiration in relation to the volume of growing media. For example:

Location and crop	Evapotranspiration	Reference
California, outdoor containers, 3.8 l of media	57–88 l, annual estimate	Furuta *et al.* (1977)
Florida, rooted cuttings, grown eight months under 73% shade, 3.8 l media	24–28 l, measured Jan–Aug	Fitzpatrick (1980)
GCRI, Southern England, pot *Chrysanthemums*, 1 l media	5–17 l, measured winter & summer crops, over 11 weeks	Bunt, this publication (§9.1.2)
Efford EHS, Southern England, nursery stock, overhead irrigation, containers on gravel	150 l m^{-2} week maximum, May–Sept (containers on irrigated sand beds require 50–60% of the overhead system)	Scott (1985)

Considerable variation in these values can be expected due to the local environment, the type and size of plant, etc.

Table 9.12   Suitability of water for irrigating pot plants (from Waters *et al.* 1972).

Water classification	Electrical conductance (mmho cm^{-1} at 25 °C)	Total dissolved solids (salts) (p.p.m.)	Sodium (% of total solids)	Boron (p.p.m.)
excellent	<0.25	<175	<20	<0.33
good	0.25–0.75	175–525	20–40	0.33–0.67
permissible	0.75–2.0	525–1400	40–60	0.67–1.00
doubtful	2.0–3.0	1400–2100	60–80	1.00–1.25
unsuitable	>3.0	>2100	>80	>1.25

9.5.2   Water quality

Irrigation waters invariably contain some salts, varying from very low concentrations, i.e. an EC of <0.25 mS cm^{-1} (= <0.25 mmho cm^{-1}) to high concentrations, i.e. EC >2.0 mS cm^{-1} (>2.0 mmho cm^{-1}). Ions which cause most problems to plants are: Na^+, Cl^-, HCO_3^- and B^{3+}; the effects which they have on plants have received special study in the Netherlands and the USA.

Gabriels (1978) distinguishes between the quality of water acceptable for continuous and short-term use. The recommended maximum levels of elements for continuous use include:

120 mg l^{-1} Ca, 25 mg l^{-1} Mg, 70 mg l^{-1} Cl, 0.5–0.75 mg l^{-1} B, 0.005–0.01 mg l^{-1} Cd, 0.2 mg l^{-1} Cu, 1.0 mg l^{-1} F, 2.0 mg l^{-1} Fe, 0.05–2.0 mg l^{-1} Mn, total hardness 350 mg l^{-1} $CaCO_3$, temporary (bicarbonate) hardness 200 mg l^{-1} $CaCO_3$ and salinity 0.85 mmho cm^{-1}.

A general classification of water quality for irrigating plants is given in Table 9.12.

9.5.3   Acidification of water

The pH of water freshly drawn from a borehole does not give a direct indication of either the amount of bicarbonates present, and the quantity of acid required for their neutralization, or the amount of free CO_2 in the water. Most of the free CO_2, i.e. dissolved CO_2 which is not combined with carbonates or bicarbonates, will be released after the water has been exposed to air, thereby causing the pH of the water to rise.

Many waters that are otherwise suitable for irrigation are high in bicarbonates of sodium, calcium or magnesium. Bicarbonates do not contribute to plant nutrition and have several adverse effects, principally that of raising the pH of the media and thereby reducing the uptake of iron. The amounts of bicarbonates in the water can be reduced by adding an acid, e.g. in the USA phosphoric acid is generally used. In the UK the advent of nutrient film and rockwool systems of culture, which use dosing and monitoring equipment, has resulted in both phosphoric and nitric acids being used to treat the water.

The amount of acid required varies with the water quality and the strength of acid. In the USA the calculated amount of acid is based upon the milliequivalents of bicarbonates present and an 'acid factor'. The acid factors, which are the amounts of acid required to neutralize 1 meq l^{-1} bicarbonate, recommended by the Soil and Plant Laboratory, California (1982) are:

Acid	Specific gravity	Acid factor (fl oz per 1000 US gal)
61.4% nitric	1.381	10.5
75% phosphoric	1.579	7.0
85% phosphoric	1.689	6.2
93% sulphuric	1.835	3.2

To adjust the water pH to approximately 6.5, add the meq of bicarbonate and carbonate in the untreated water, subtract 2.0 meq and multiply the remainder by the acid factor. For example, 5.5 meq − 2.0 = 3.5 meq × 7.0 = 24.5 fl oz of 75% phosphoric acid per 1000 gal. If the total milliequivalents of carbonates are multiplied by the acid factor, the pH of the water would be approximately 4.3.

In the UK the amount of acid required is based on titrating the water with the desired acid. Figure 9.5 shows titration curves for three water supplies having total alkalinity of 151, 286 and 396 mg l^{-1} calcium carbonate equivalent. The quantity of nitric acid (70% w/w) required to reduce the pH to 5.8 ranges from 95 to 245 cm^3 per 1000 l water. When calculating the nutrient content of a liquid feed an allowance must be made for the amounts of nitrogen or phosphorus supplied by the acids. Each 100 cm^3 of 70% nitric acid added to 1000 l of water supplies 22 mg l^{-1} of nitrogen; and 100 cm^3 of 75% phosphoric acid in 1000 l of water supplies 37 mg l^{-1} of phosphorus. The amounts of nitrogen and phosphorus supplied by nitric and phosphoric acids of varying specific gravities and purities can be calculated from:

$$\text{acid volume} \times \text{specific gravity} \times \frac{\% \text{ acid}}{100} \times \text{factor}$$

The factor for nitric acid is 0.222, and for phosphoric acid 0.316.

At the Efford Experimental Horticulture Station (UK) acidification of the water (pH 7.8–8.3, 225 p.p.m. calcium carbonate equivalent) to pH 5.8 with nitric acid has given marked improvements in growth and quality of several hardy ornamental nursery stock species. As well as calcifuge species, e.g. *Azalea, Erica, Magnolia, Picea* and *Skimmia*, lime-tolerant species such as *Acer, Ceanothus, Choisya, Cytisus* and *Cornus* have also responded.

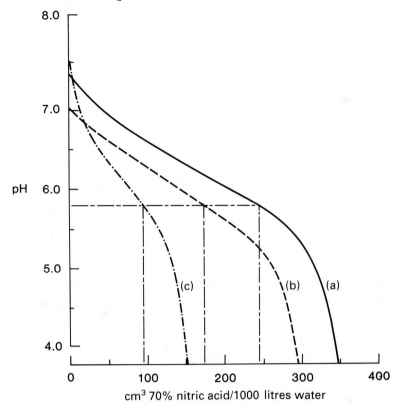

Figure 9.5 Titration curves of three water supplies with hardness equivalent to calcium carbonate at rates of: A 396 mg l^{-1}, B 286 mg l^{-1}, C 151 mg l^{-1}. The amounts of nitric acid required to reduce the pH to 5.8 are shown.

Acids can be added using either twin-headed injectors or dosing equipment; because of corrosion, especially with nitric acid, the equipment should be made of stainless steel or PVC.

9.5.4 Water purification

Surface drainage water drawn from ponds and rivers may contain bacteria, e.g. *Erwinia*, and fungi, e.g. *Pythium* and *Phytophthora*, which can cause plant losses. Water drawn from the public supply and boreholes is usually free of plant pathogens.

Chlorination The most practical way of purifying contaminated water is by chlorination. When a chlorination compound such as sodium hypochlorite or chloroisocyanurate, e.g. 'Fi-chlor Clearon' tablets, is added to water, some of the chlorine destroys any organic matter and ammoniacal nitrogen there may be in the water. The excess of chlorine is known as 'free' chlorine and can be measured colorimetrically. The amount of free

227

chlorine required to kill fungal spores and bacteria depends upon the temperature of the water and the time that the free chlorine is in contact with the pathogen before the water is applied to the plants. Spores of *Phytophthora cinnamomi* are killed by a concentration of 2 p.p.m. of free chlorine with a minimum contact time of 1 minute. Plant species suscept-ible to the disease, e.g. *Calluna, Azalea, Chamaecyparis* and *Thuja*, have not been damaged by 10 p.p.m. of free chlorine (Scott *et al.* 1984). Fi-clor Clearon has been safer than sodium hypochlorite but is more expensive. Before treatment the water should first be passed through a sand bed to remove algae, soil particles, etc.

A cheaper method of chlorinating water is to inject chlorine gas (Daughtry 1983). This forms hydrochloric and hypochlorous acids in the water, but the concentration of hydrochloric acid is too low to have any effect on the pH. The hypochlorous acid dissociates to form hydrogen and hypochlorite ions. However, *handling chlorine gas is potentially very danger-ous*, for safety and to maintain control over the concentration of free chlorine a vacuum injector is used.

Ultraviolet radiation   Bacteria present in the water can also be destroyed by ultraviolet light at a wavelength of 265 nm. However, ultraviolet radiation has a very short penetration in water, and the depth of treated water should therefore not exceed 10 cm.

Reverse osmosis   This method is primarily used to remove high concen-trations of salts, the water being forced across a membrane in the reverse direction to osmotic diffusion. The membranes can be of cellulose acetate or polyamide and remove about 85–95% of the dissolved inorganic salts. The method is used in Holland and parts of the USA where Na^+ and Cl^- ion concentrations are too high for good plant growth. Only about 50% of the original volume of water is recovered in the process.

Demineralization or Deionization   Water is passed through a column containing anion and cation exchange resins. Depending upon the resins used, water of high purity can be produced. When the conductivity of the treated water starts to rise, indicating that the resins are approaching saturation, they are recharged for further use. Demineralization is gener-ally too expensive for nursery use, except for producing boiler feed water. Domestic-type water softeners which replace the calcium and magnesium in the water with sodium are not suitable, as the sodium is more harmful than the calcium and magnesium that have been replaced.

CHAPTER TEN

Irrigation systems

Plants grown in pots have only relatively small reserves of water in the media from which to meet their requirements, and these soon become exhausted unless replenished by frequent irrigations. Some plants may require as many as three irrigations per day in mid-summer to prevent wilting. The high cost of labour has encouraged the use of various automatic or semi-automatic systems of irrigation and the majority of commercially grown pot plants are now watered by one or other of these systems. In the following discussion of irrigation systems, it is not the intention to make a critical appraisal of proprietary equipment, but rather to discuss the general principles and characteristics of each system in so far as they interact with the media and affect nutrition and management. Irrigation systems can be classified into four groups:

(a) drip and trickle systems,
(b) capillary watering,
(c) flooded benches,
(d) overhead sprinklers.

10.1 DRIP SYSTEM

With this system a plastic hose of 1.25–2 cm ($\frac{1}{2}$–$\frac{3}{4}$ in) diameter is laid along the centre of the bench. From this hose small-bore capillary tubes, sometimes known as 'spaghetti' lines, are taken to each pot and held in position over the surface of the medium with either a peg or a weight. The rate at which water flows from the tubes is determined by the water pressure in the distribution mains and by the diameter and length of the small-bore tubes. The tubes are usually of 0.75 mm ($\frac{1}{32}$ in) internal diameter. If tubes with a larger internal diameter are used, they have a restriction or nozzle fitted at the discharge end to control the rate of flow; this can be removed for cleaning if a blockage occurs. Usually the rate of water flow is between 1 and 2.25 l h^{-1} (1.5–4 pints h^{-1}). This is slow enough for the water to drip from the end of the tubes, hence the term 'drip system'. The frequency and duration of the irrigation can be controlled either manually or automatically by the use of time switches or conductivity meters wired to solenoid valves.

From the cultural viewpoint the advantages of the drip system of irrigation are:

(a) The ability to control the soil moisture tension and thereby exercise some control over the rate of growth. Plants can be grown relatively wet, to obtain the maximum growth rates, and then the water content can be reduced, to produce a check to vegetative growth and so encourage a more reproductive condition, e.g. the even production of winter flowering in the pot chrysanthemum.
(b) Pots can be leached as often as required to prevent a build-up of salts in the mix.

The disadvantages of this system are:

(a) All plants receive the same amount of water and nutrients irrespective of any difference in their growth rates and water requirements. Variations that naturally occur between plants in their water requirement mean that some plants could be over-watered while others are under-watered.
(b) Because of the low rate of water application the lateral spread of water in some mixes can be inadequate; one part of the pot can be dry while another part is wet. This situation is worse when the 'plant' consists of several rooted cuttings grown in the same pot, e.g. the pot chrysanthemum. It is most likely to occur when the mix has been allowed to become too dry between irrigations.
(c) Excessive amounts of water can easily be given by leaving the system turned on too long or by using it too frequently. It is desirable, therefore, to choose a medium that drains freely and always has an acceptable level of aeration; this offers some safeguard against over-watering. The detrimental effects of excessive water application and the consequent lack of aeration on plant growth has already been discussed (§ 3.4).

10.1.1 Blocking of nozzles

Small-bore irrigation tubes or nozzles are subject to blocking due to:

(a) solid particles in suspension,
(b) growth of bacteria and algae,
(c) chemical precipitates.

Prevention or remedial measures include:

(a) Solid particles should be removed by filtration, e.g. fitting 200-mesh filters or using sand filters.
(b) Blockages by bacterial and algal slime can be prevented by treating static water tanks with an algicide, e.g. 'Dichlorophen', or by periodically injecting hydrogen peroxide into the lines at a dilution of 1 in 150;

leaving in the lines for two hours or overnight. *Care is required in handling and storing hydrogen peroxide.* The water supply can also be chlorinated to give a residual free chlorine concentration of 0.5–1.0 p.p.m. with the pH at 7.0. Alternatively, where iron bacteria occur, periodic treatment of the water to give 50 p.p.m. of free chlorine, with a contact time of 30 min before flushing out the system, should be carried out.

(c) Chemical precipitates can be removed by acid treatment. Normally nitric acid (70%) is injected into the lines at a dilution of 1 in 400, left overnight and then flushed out. It is sometimes necessary to pre-dilute the acid to, say, one quarter strength before making the final dilution. Hydrochloric and sulphuric acids have also been used; *because of the corrosive risk and danger in handling acids, advice should first be obtained from the advisory officer or extension agent.*

Iron in the water supply can be particularly troublesome in causing blockages and this may require a combination of precautions and remedial treatments. It is advisable not to use water having more than 0.5 p.p.m. of iron and 2 p.p.m. of total sulphides.

10.2   CAPILLARY WATERING

With the capillary system of watering, plants are stood upon a layer of wet sand or matting. The water flows by capillary attraction through the base of the pot into the medium. Provided that the sand or matting is kept wet and the capillary action is not broken, no further watering is required.

This system has the advantages of:

(a) reduced labour, especially with small pots;
(b) a relatively constant water tension is maintained and very wet conditions with low air-filled porosities are avoided;
(c) irrigation of plants with a rosette habit or which are sensitive to foliar splashing, e.g. *Saintpaulia*, is easier.

10.2.1   *Sand benches and beds*

In the system developed by the National Institute of Agricultural Engineering (Wells & Soffe 1962) a constant water table is maintained 5 cm (2 in) below a layer of sand which must have a particle size distribution that gives good capillary conductivity. Specific disadvantages of the system are: the benches must be level within ± 3 mm ($\frac{1}{8}$ in) and they must be strong enough to support the weight of the wet sand.

A similar system is used for nursery stock containers, with the beds being at ground level. The sand particle grading should be:

$$30\text{–}45\% \; >0.5 \text{ mm}$$
$$40\text{–}60\% \; 0.2\text{–}0.5 \text{ mm}$$
$$5\text{–}15\% \; <0.2 \text{ mm}$$

As well as supplying water to the container by capillary action during hot dry periods, the reverse action occurs after heavy rainfall. Water is drawn out of the container, improving the aeration of the mix and giving better quality with plants such as *Erica* which are susceptible to waterlogging over winter.

One modification of the sand bed system is to periodically irrigate the sand with a seep hose or drip lines instead of having a controlled water table.

10.2.2   Capillary matting

To avoid the high costs of installing sand benches in greenhouses and to obtain greater flexibility of management, capillary mats of synthetic non-woven felts are commonly used. Verwer (1978) examined a number of materials for their water-storage capacity, capillary conductivity, resistance to shrinking and decomposition, pH and salt content. Mats made of polyester with a water-holding capacity of $4 \, l \, m^{-2}$ and a capillary rise of about 12–20 cm, as defined by the Federal German Republic Standard DIN53924, are preferred to those having some wool or cotton. This system of watering is not generally used with containers more than 15 cm (6 in) deep.

10.2.3   Algal control

The high moisture level in the sand or capillary matting is conducive to the growth of algae, especially if nitrogen and phosphorus are added to the irrigation water. Algal growth can be controlled by:

(a) steam sterilization of the matting between crops,
(b) allow matting to dry out and brush with a stiff broom,
(c) using algaecides.

To be successful the algaecide must be non-phytotoxic to the crop as well as able to control algal growth. Examples of chemicals that meet these criteria are algamine ('Killgerm') and dichlorophen ('Panacide') in the UK, and 'Cyprex 65W' (with a spreader-sticker such as 'Exhalt 800' added) in the USA. It is always advisable to first test the algaecide with the crop under the environmental conditions which are to be used. Algal growth on the surface of growing media can be controlled by applying a 'Mancozeb' solution to wet the top 2 cm (1 in) of media. Some algaecides, e.g. 'Gloquat-C', will also prevent the roots of many container plants growing into the sand or the capillary mat.

A recent alternative method of controlling algae has been to cover the matting with a dark, perforated polyethylene sheet (Bjerre 1983). As well as excluding light from the matting, the covering reduces water evaporation to one third of the rate of uncovered matting and also raises the temperature within the pots by 1 °C.

Table 10.1 Air–water relations of three mixes irrigated by different systems.

Watering system	Mix		
	1	2	3
hand watered (drained for 8 h):			
air-filled porosity (%)	5.4	8.0	19.1
water (%)	75.6	63.1	52.9
total pore space (%)	81.0	71.1	72.0
capillary watered (pots 5 cm above the water table):			
air-filled porosity (%)	17.1	18.0	36.6
water (%)	63.6	53.5	34.5
total pore space (%)	80.7	71.5	71.1

10.2.4 Physical and nutritional effects

The principal physical effect of capillary irrigation on the medium and plant is the maintenance of a near constant matric tension. Any slight increase in tension during periods of high evapotranspiration is soon restored to normal, i.e. about 10–15 cm water tension. Although the tension is low, it is sufficient to maintain adequate air-filled porosities with many media. Examples of this effect with three mixes are shown in Table 10.1. The system does not, however, allow plants to be grown 'dry' for hardening off, etc.

Capillary irrigation has two main nutritional effects. The near constant water content means that in the short term the osmotic stress is also constant; this contrasts with containers that are allowed to become dry, which then develop large osmotic stresses. In the absence of leaching, the upward flow of water and liquid feed into the container results in an increase in salinity if more nutrients are being supplied than are absorbed by the plant. Guttormsen (1969) also showed that there was a very strong salinity gradient within the pot. With a feed having 200 p.p.m. of nitrogen and potassium, salinity in the lower part of the pot was 2 mS cm^{-1}, whereas in the top of the pot it was 9 mS cm^{-1}. The strength of the feed needs to be adjusted to the crop and amount of water being lost by evapotranspiration. In practice feeds of about 100 p.p.m. nitrogen, 15 p.p.m. phosphorus and 100 p.p.m. potassium are used. If salts do accumulate in the pot a heavy surface watering will correct the situation.

10.3 FLOODED BENCHES

With the original system of flooded benches developed at Cornell University (Seeley 1947), the pots were stood on a watertight bench which could be flooded, usually until the pots were $\frac{1}{3}$–$\frac{1}{2}$ submerged, although sometimes a greater depth of water was necessary. When the soil surface appeared to be moistened the water was drained from the bench into a reservoir for recirculation. More recent designs of flooded benches have channels to distribute the water quickly and evenly to a depth of 1–2 cm. As with capillary irrigation, the flow of water into the container is upward; apart from the shallow layer in the base of the pot, the medium is not leached. If the medium does not have sufficiently large pores, root development in the base of the pot can be adversely affected.

10.4   OVERHEAD SPRINKLERS

Although used only for a few crops in greenhouses, e.g. lettuce grown in beds, this system is widely used for irrigating container-grown nursery stock. It has a lower installation cost than drip systems but the water use efficiency, i.e. the percentage of applied water retained in the container, is also much lower. Factors affecting the water use efficiency with overhead sprinklers are: the density of the containers and the amount of water that is able to reach the container through the foliage. In a two year study Weatherspoon & Harrell (1980) found that nearly three times as much water was applied by overhead sprinklers as with a drip system.

Often a larger surface area of the mix will be wetted with this system than with a drip irrigation, this reduces the risk of dry areas developing in large containers. Leaching can occur as with all surface systems of irrigation.

CHAPTER ELEVEN

John Innes composts

The John Innes composts were developed as a result of investigations by Lawrence & Newell (1939), at the John Innes Horticultural Institution, into the high mortality rate of *Primula sinensis* seedlings being raised for genetical studies. This work was the first attempt by researchers to produce a standardized medium, suitable for a wide range of plants. Prior to this, there had been no standardization of materials or compost formulae; growers prepared individual mixtures for each species of plant from carefully preserved and guarded recipes. In addition to the well-rotted turf loam, chopped with a spade into walnut size pieces, which formed the basis of the composts, a wide range of other materials was used, including mortar rubble, crushed brick, burnt clay, charcoal, well-rotted animal manure, spent hops, leaf mould, etc., and fertilizers such as bone meal, steamed bone flour, flue dust, bonfire ashes, etc.

The John Innes composts differed from the traditional composts in a number of important ways:

(a) a single physical mixture was found to be suitable for a wide range of plants;

(b) they were based upon steam sterilization of the soil which, if done correctly, eliminated the heavy losses of seedlings resulting from soil-borne fungal and insect attacks;

(c) a 'sterile' and relatively standard organic physical conditioning material, i.e. peat, was used, in place of rotted animal manure and other materials of variable physical and nutritional properties;

(d) fertilizers were added with precision on a weight basis, in place of the traditional advice of '. . . to each barrow-full of this mixture add a six inch pot-full of bone meal, sprinkle a handful of lime and . . .'.

11.1 FORMULATION

Essentially, only two John Innes composts were formulated; one with a low nutrient content was used for seed sowing and the rooting of cuttings; the other was a potting compost for general pot work. To allow for the varying degree of plant vigour and for seasonal changes in the growth rates, the nitrogen, phosphorus and potassium levels in the potting compost were adjusted by varying the rate of application of a 'complete' base fertilizer containing these three elements. The formulae of the composts are as follows:

	Bulk ingredients (by volume)	Fertilizers	$lb\,yd^{-3}$	$kg\,m^{-3}$
Seed Compost	2-loam	superphosphate	2	1.186
	1-peat	calcium carbonate	1	0.593
	1-sand			
Potting Compost	7-loam	hoof and horn	2	1.186
(JI Potting	3-peat	superphosphate	2	1.186
Compost no. 1	2-sand	potassium sulphate	1	0.593
(JIP-1))		calcium carbonate	1	0.593

Although the term 'sand' is conventionally used, the particles grade up to 3 mm ($\frac{1}{8}$ in) diameter and the term 'grit' would be more correct. It is possible to purchase the hoof and horn, superphosphate and potassium sulphate as a complete ready-mixed fertilizer, it is then known as the 'John Innes Base' fertilizer and has an analysis of 5.1% N, 7.2% P_2O_5, 9.7% K_2O.

The rates of added nutrients ($mg\,l^{-1}$) for the JIP-1 compost are:

N	P	K
154	106	243

Where higher nutrient levels are required, the rates of hoof and horn, superphosphate and potassium sulphate (or, for convenience, the JI Base fertilizer) are doubled to give the JI Potting Compost no. 2 (JIP-2), or trebled to give the JI Potting Compost no. 3 (JIP-3); the rates of calcium carbonate are also increased accordingly.

11.2   COMPOST INGREDIENTS: LOAM

Loam is the most important of the ingredients in the JI composts. Apart from its effect upon the physical and biological properties of the compost, it forms the basis of plant nutrition by supplying clay, which has a cation and anion exchange capacity, microelements and organic matter, which gives a slow release of nitrogen to the plants. Loam is also the most variable of the ingredients in the compost, both physically and chemically, and careful attention must therefore be given to its selection and preparation, including steam sterilization, if good results are to be consistently obtained.

11.2.1   Selecting a loam

The most important factors to consider when selecting a loam are: pH, texture, structure, organic matter, and response to steam sterilization.

pH   Loams chosen for making the John Innes composts should have a pH within the range 5.5–6.5. The pH of the loam is important for two reasons: its effect upon the mineral nutrition of plants, and its effect on the reaction of the loam to steam sterilization. When loams have a pH below 5.5, there is the risk of a low level of fertility; the rate of phosphorus fixation may be high and there is also the possibility of a large release of manganese following steam sterilization. If the loam has a pH above 6.5, there is the possibility of microelement deficiencies such as boron and iron and the very high risk of nitrogen toxicities developing after steam sterilization.

Acid loams must be limed to bring the pH up to 6.3. If the actual amount of lime required to raise the pH to this value is not known, an estimate of the amount of calcium carbonate required can be made using the following formula:

desired increase in pH $\times$ 2.67 = kg calcium carbonate per cubic metre, or

desired increase in pH $\times$ 4.5 = lb calcium carbonate per cubic yard

e.g. for a loam with a pH of 5.6,

$(6.3 - 5.6) \times 2.67 = 1.869$ kg m^{-3} calcium carbonate

The calcium carbonate is added to alternate layers of turfs when making the loam stack (§ 11.2.2).

Although the actual lime requirement for a given change in pH will depend upon a number of factors, of which the clay and the organic matter contents are the most important, the above formula has been found to give reasonable estimates of the lime requirement of soils having about 20% clay content. Loams with a pH above 6.3 are better avoided and under no circumstances should loams having a pH of 7 or more, with free carbonates present, be used for compost making.

The effect of the pH on the biological and chemical changes induced by the heat sterilization of the soil are not always fully appreciated. The risk of manganese toxicity occurring if the pH is low has to be balanced against ammonium toxicities occurring at high pH values; unless great care is exercised, it is possible to move away from one form of toxicity into another. The importance of adequate pH control of the soil cannot be overemphasized. Manganese toxicity is discussed more fully in Chapter 12 and some of the interactions which occur between steaming, liming and organic forms of nitrogen are described in section 11.5 and Figure 11.2.

Texture   This term refers to the composition of the soil, i.e. the amounts of clay, silt and sand as determined by a mechanical analysis. Clay loams,

237

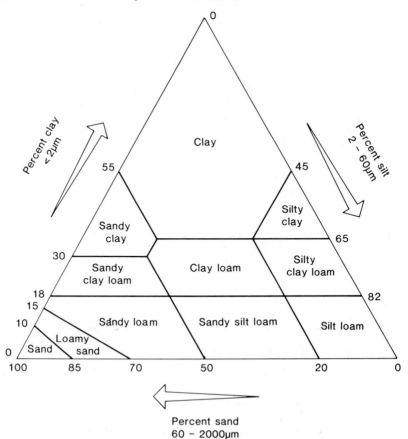

Figure 11.1 Classification of soil texture based on particle size grades adopted by the British Standards Institute and the Massachusetts Institute of Technology.

i.e. those with a 20–35% clay, are preferred for making John Innes composts (Fig. 11.1). This classification of soils is based on the particle size grades adopted by the British Standards Institute and the Massachusetts Institute of Technology (Table 11.1).

A clay loam can be identified by its ability, when moist, to take on a smear or polished surface when rubbed between the fingers. Sand gives a gritty or rough feel, and silt has a smooth or silky feel but does not give a polished surface.

Structure This refers to the grouping of the particles of clay, silt, sand and organic matter into small aggregates. It is also desirable that these aggregates have a good stability and do not breakdown and disperse when the soil is wet. It has long been known that grass roots improve the soil structure; a permanent ley has a much better structure and aggregate stability than an arable soil of similar texture. Alvey (1961) studied the way in which growing four grass species affected the aggregate stability

238

Table 11.1 Particle size fractions on which the mechanical analysis of soil is based.

		Particle size (mm)
clay		<0.002
silt	fine	0.002–0.006
	medium	0.006–0.02
	coarse	0.02–0.06
sand	fine	0.06–0.2
	medium	0.2–0.6
	coarse	0.6–2.0

of three arable soils having clay contents ranging from 18 to 29%. On each of the soils, separate plots were sown with either rye grass, cocksfoot, fescue or timothy, and an additional plot was maintained in a fallow condition. Each treatment was assessed annually over a four year period for its effect on the mechanical stability of the soil aggregates which was measured by a wet sieving technique. With all three soils, the grass treatments showed an increase in the aggregate stability by comparison with the plots maintained in a fallow condition. When the soils were used to make John Innes composts, the grass treatments also showed a greater release of mineral nitrogen; this was derived from their higher organic matter contents.

Organic matter In addition to its indirect effect in helping to improve soil structure, organic matter also provides a supply of plant nutrients, principally of nitrogen. The organic matter should be present partly as fibrous, relatively undecomposed grass roots and partly in the more decomposed form as humus. A loam with a good organic matter content will give a steady release of nitrogen over a long period. For this reason leys are preferable to arable soils for compost making.

Response to steam sterilization Steam sterilization of the loam to eliminate the fungi and insects, which would otherwise attack the seedlings, is an essential part of the preparation of the John Innes composts. Unfortunately, in addition to eliminating the pathogens, the steam sterilization of some soils can lead to the production of substances harmful to plant growth. These toxins can be of mineral or organic origin and, although the risk of toxicity can be minimized by avoiding loams with either a high or low pH and those with a high organic matter content, the effect of steam sterilization on subsequent seedling growth cannot be precisely predicted from either a visual examination of the loam or its chemical analysis. Whenever a new loam is to be used for compost making, it is prudent to first make a small growing test, using seedlings such as antirrhinum, to compare growth in the steam sterilized new loam with the growth of similar seedlings in another loam whose reaction to sterilizing is already known. If growth in the new loam is comparable to that in the other soil and no toxicities are apparent, it is then safe to proceed.

11.2.2 Making a loam stack

Having selected a loam which meets the above requirements, it should then be made into a loam stack; this presents an opportunity of adjusting its pH level and increasing its general fertility. Turfs are cut to a depth of 10 cm (4 in) in the spring of the year when the loam is still moist and the growth is active. A loam stack about 2 m square (6 ft square) is made by first forming a layer of loosely packed turfs, grass-side facing downwards; upon this is placed a 5 cm (2 in) layer of strawy animal manure, which is followed by another layer of turfs.

Adjustments for low nutrient levels   Loams vary in their general fertility and should be analysed for available levels of phosphorus and potassium, any deficiency being corrected by adding superphosphate and potassium sulphate while making the loam stack. For economic reasons, stacking the loam as detailed above is not always practised, any deficiencies of phosphorus and potassium must then be corrected by adding extra fertilizers to the compost. The rates given below are based on analyses of the compost, the extractants being sodium bicarbonate at pH 8.5 for phosphorus, and 1 M ammonium nitrate for potassium and magnesium.

nutrient index	phosphorus		potassium		magnesium	
	0–1	2–4	0–1	2–4	0–1	2–4
mg l^{-1}	0–15	16–70	0–120	121–600	0–50	51–250

Extra fertilizer rates (kg m^{-3})

	superphosphate		potassium sulphate		Kieserite	
JIP-1	1.9	1.1	0.7	0.6	1.5	1.1
JIP-2	3.8	2.3	1.5	1.2	3.1	2.3

If the pH is below 6.3, lime is added according to the formula given in section 11.2.1, the lime and the strawy animal manure being added alternatively between the layers of turfs. Also, if the loam is in a dry condition when the heap is being made, it should be watered generously, otherwise the animal manure and the grass roots will not decompose sufficiently. When the loam stack cannot be built under cover, some protective covering must be erected in time for the heap to dry out before it is sterilized; wet soils cannot be sterilized successfully. The loam will be ready for use in about six months after stacking. It is prepared for sterilizing by first chopping it down in thin layers to minimize any variation within the heap. After this it is put through a mechanical shredder or a hand sieve.

11.3   PEAT

The function of peat in the John Innes composts is primarily that of a physical conditioner. It increases the total porosity and improves both the water retaining capacity and the aeration. Either sphagnum or sedge peat having particles grading evenly up to 10 mm ($\frac{3}{8}$ in) can be used, but fine dusty peats and so-called 'rhododendron' peats with a low pH are not suitable. The pH of the peat should be between 3.5 and 5.0 and, unlike loamless mixes where the peat usually forms the greater part of the bulk, additional lime to correct for the acidity of the peat is not required; liming the loam to pH 6.3 is sufficient. The small quantity of calcium carbonate included with the base fertilizer serves only to neutralize the acidity developed by the fertilizers. The peat is not normally sterilized but it must be moistened before being used; air-dry peat does not mix well with the loam and the sand.

11.4   SAND

Sand is used as a physical conditioner. To enable it to perform its function of draining excess water from the compost, it must have a coarse grading, i.e. 60–70% of the particles being between 1.6 and 3.2 mm ($\frac{1}{16}$–$\frac{1}{8}$ in) diameter. Technically, it would be more correct to use the term 'grit' for material of this size, but the term 'sand' has traditionally been used with reference to the John Innes composts; the true horticultural sands with fine particle sizes are not suitable for use in the John Innes composts. Apart from its particle size, it is also important to ensure that the sand is free from carbonates; the use of sand containing carbonates can result in a high pH with all its associated nutritional problems.

11.5   PASTEURIZATION

To make a John Innes compost *it is essential that the loam is pasteurized* and, provided it is done efficiently, heat pasteurization, by means of steam or electricity treatment, has given better results than with chemicals. Two types of toxicity can occur as a result of pasteurizing the soil; manganese toxicity is associated with some types of soils having a low pH, and nitrogen toxicity may occur when soils rich in organic matter and having a high pH are steamed. Pasteurization speeds up the biological break-down of organic matter and also introduces a temporary blockage in the nitrogen cycle. Those micro-organisms in the soil which convert ammonium to nitrites and then into nitrates are eliminated by steaming and, for a period, their re-establishment is inhibited, thereby resulting in a build-up of ammonium which may reach toxic levels.

An example of the interactions which can occur between heat treatment, the soil pH and the amount of readily nitrifiable organic matter is shown in Figure 11.2. Tomato seedlings were pricked-out into loam-based John Innes type composts; the unlimed soil had a pH of 5.17 and it was known from previous experience to be of a type which did not produce manganese toxicity after it had been steamed. The results showed that, when the nitrogenous fertilizer was hoof and horn, neither steaming nor liming by themselves caused any significant degree of toxicity but when the soil was limed to pH 6.3, steamed and then made into a compost, nitrogen toxicity occurred. This toxicity was proportional to the amount of readily mineralized nitrogen present (Fig. 11.3). When the nitrogen source was urea-formaldehyde, however, no toxicity developed; this was because of the slower rate of nitrogen availability. Other experiments have shown that toxicity can occur even when the amount of hoof and horn fertilizer added to the compost is low but the amount of organic matter in the loam is high, i.e. toxicities result whenever the level of readily nitrifiable organic matter is high, irrespective of whether it originates from a nitrogenous fertilizer or from using soils such as old cucumber beds which are high in organic matter. From other experimental results obtained by the author (Bunt 1956), it was found that the breakdown of organic matter and the release of ammonium is not affected by the sequence of adding the lime, i.e. the same results are obtained irrespective of whether the lime is added immediately before or after pasteurization.

Figure 11.2 The effects of steam sterilizing and liming the loam, when the compost contains a large amount of organic nitrogen. *Top left*: not steamed, not limed. *Top right*: not steamed, limed. *Bottom left*: steamed, not limed. *Bottom right*: steamed, limed.

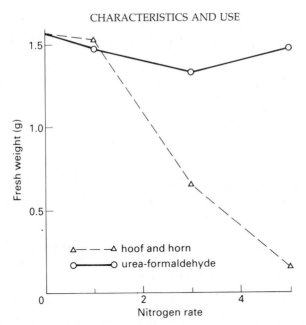

Figure 11.3 The effect of nitrogen source and rate on the growth of tomato seedlings in a loam compost. When a readily mineralized source of nitrogen such as hoof and horn was used, the depression in growth was related to the amount of nitrogen. With urea-formaldehyde there was no depression in growth even at the highest rate of nitrogen. The loam had been steamed and limed for both sets of composts. The nitrogen rates were equivalent to JIP-1, JIP-3 and JIP-5.

11.6   CHARACTERISTICS AND USE

The concentration of the principal plant nutrients in the compost, i.e. the nitrogen, phosphorus and potassium, can be adjusted by increasing the quantity of base fertilizer, and the choice of compost strength is determined by the vigour of the crop and the season. For example, primula seed is sown in JI-Seed compost and pricked-out into JIP-1 compost, whereas cucumber seed is germinated in JIP-1 and potted into JIP-2. Tomato plants are pricked-out into JIP-1 compost in winter and into JIP-2 compost in spring, and for plants such as tomato and chrysanthemum which are grown to the fruiting or flowering stage in 10 in pots, the JIP-3 compost is used.

11.6.1   Nutrient levels

Increasing the nutrient strength of the compost does not necessarily extend the period for which plants can be grown before feeding is required. Provided that the light intensity and temperature are not limiting, plants grown in composts with high nutrient levels have the highest growth rates. If the plants are not given supplementary liquid feeding however, rates of dry matter production and the percentage of nitrogen in the tissues decline at similar rates in all composts (Bunt 1963).

11.6.2 Storage

If composts are kept for a period before they are used, several biological and chemical changes can occur. Apart from the levels of exchangeable manganese (§ 12.8.3), the most important changes are the forms and amounts of mineral nitrogen and the compost pH.

In composts made with steamed loam there is a temporary inhibition of the nitrifying bacteria, but not of the ammonifiers. The ammoniacal nitrogen produced from the organic matter in the loam and the fertilizer causes the pH to rise, by about 0.5 of a unit in the example shown in Figure 11.4. When nitrification commences the pH falls, and can drop to one unit below the original value. If the loam has not been heat pasteurized ammonium nitrogen does not accumulate, being rapidly converted to nitrate nitrogen, and there is a progressive fall in the compost pH.

The salinity is not directly affected by the increase in ammonium

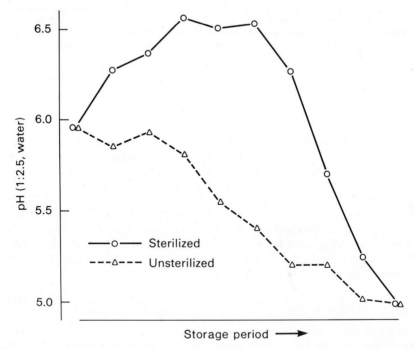

Figure 11.4 Changes occurring in the pH of a John Innes compost (JIP-1) during storage. The rate of change will be dependent upon the amount of organic matter, the form and amount of nitrogenous fertilizer, the temperature and moisture content of the compost (from Bunt 1956).

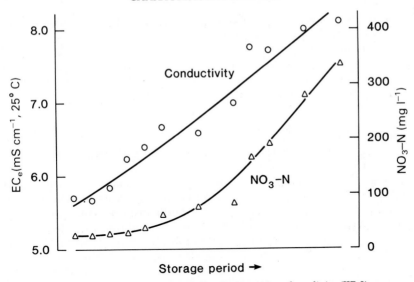

Figure 11.5 Nitrate formation during storage raises the salinity (JIP-3).

nitrogen, but there is a significant rise as nitrates are formed (Fig. 11.5). Typical changes which can occur in the salinity values of the JI composts are:

Compost	EC_e (mS cm^{-1} at 25 °C)	
	freshly mixed	after storage and nitrate formation
JIP-1	3.69	5.36
JIP-2	4.56	6.84
JIP-3	5.58	8.10

The extent of these changes in pH and salinity depends upon the amount of organic nitrogen that can be mineralized. At equivalent rates of nitrogen, urea-formaldehyde does not mineralize as quickly as hoof and horn; the potential chemical changes during storage and any subsequent check to seedling growth with this fertilizer will therefore be less than with hoof and horn (see Fig. 11.3). In other experiments comparing fresh and stored composts made with various inorganic and slow-release forms of nitrogen, the vigour of seedling growth was related to the salinity of the compost (see also section 8.14.1 for depression of seedling growth and salinity). Thus, the extent to which organic or slow-release forms of nitrogen have mineralized during storage is more important than the age of the compost *per se*.

The *rates* at which these changes occur depends primarily upon the moisture content of the compost and the temperature at which it is stored.

In composts that were kept dry and at low temperatures nitrification did not commence for several months.

The chemical changes that occur in loam composts during storage also apply to peat mixes when organic type nitrogenous fertilizers are used. However, high levels of ammonium are more phytotoxic in peat mixes than they are in loam-based mixes (Tew Schrock & Goldsberry 1982), this is in part due to the lower pH and slower rate of nitrification in peat mixes.

11.7 COMPOSTS FOR CALCIFUGE PLANTS (JIS (A))

It is well known that calcifuge plants, such as *Erica* and *Rhododendron*, do not thrive in composts which have a pH approaching the neutral point; often the plants turn chlorotic and make little growth. The problem was investigated by Alvey (1955) in relation to the John Innes composts. *Erica gracilis* was grown in both the standard John Innes seed compost and the potting compost (JIP-1). These composts were also modified by replacing the 0.6 kg m^{-3} (1 lb yd^{-3}) of calcium carbonate, normally added with the base fertilizer, with flowers of sulphur at either 0.6 or 1.2 kg m^{-3} (1 or 2 lb yd^{-3}). Other treatments included a ferrous sulphate solution at 6 g l^{-1} (1 oz gal^{-1}) watered on to the compost, and foliar sprays of ferrous and manganese sulphates. Although the amount of growth in the standard JI-Seed and JIP-1 composts was good by comparison with a typical erica compost, the incidence of chlorosis was high, especially in the JIP-1 compost. In composts where the calcium carbonate had been replaced with sulphur, there was good growth and chlorosis was virtually eliminated. The fresh weights, chlorosis rating and pH of the composts are given in Table 11.2. At the highest rate of sulphur, there was a depression in growth; for this reason, the seed compost with the lower rate of sulphur was chosen as being the best compost for growing ericaceous plants and was described as JIS(A), the (A) denoting the non-standard, acid compost.

Bunt (1956) examined the rate at which sulphur was oxidized in composts and found that, under typical glasshouse conditions of high soil

Table 11.2 Growth of *Erica gracilis* in different composts.

Compost	Fresh weight (g)	Chlorosis rating	pH at potting	pH after cutting
JIS	1490	53	6.4	6.6
JIS + 0.6 kg m^{-3} sulphur	1604	8	5.7	4.7
JIS + 1.2 kg m^{-3} sulphur	751	0	5.7	3.9
JIP-1	1021	139	6.5	6.1
JIP-1 + 0.6 kg m^{-3} sulphur	1719	53	5.7	5.0
JIP-1 + 1.2 kg m^{-3} sulphur	1483	0	5.8	4.2
'Grower's'	907	30	4.4	4.9

temperatures and moisture levels, the maximum pH reduction in the compost occurred within 6–8 weeks.

11.8 OTHER LOAM-BASED MIXES

Many growers recognize the advantages of having some loam in their potting mixes, primarily to provide more buffering against the effects of rapid depletion of nutrients as occurs in peat and lightweight mixes. To combine the advantages of easier management of nutrition in loam-based mixes with the better physical properties of lightweight mixes, some growers use equal parts of loam, peat and a lightweight aggregate such as perlite. Although the mineral soil content has thereby been reduced from the 58% in the JI composts to 33%, there is still sufficient soil for the mix to be considered as being soil-based. However, if the soil content of the mix is reduced to only 10%, management can become more difficult. For example, peat-based mixes are normally used at pH values one unit below that of loam composts, at such acidity levels there can be a substantial release of manganese from the steamed loam. Also, with long term cropping the microelement levels in the mix can become limiting unless fritted trace elements are also supplied.

CHAPTER TWELVE

Heat pasteurization

Practically all of the soil-borne pests and diseases which attack seedlings and pot-grown plants can be eradicated by heat treatment of the materials before they are used for making potting media. This method of controlling soil pathogens was first used by glasshouse growers at the end of the nineteenth century and rapidly increased in importance until the 1950s, when virtually all the mineral soil used in potting media was heat treated. Since this time, the increasing use of loamless mixes made from materials such as peat, perlite, vermiculite, plastics, etc., which are virtually free of pathogens and therefore do not require treatment, has meant that heat pasteurization is not such an important factor in the preparation of potting media as it once was. In the UK the term 'sterilization' has traditionally been used to describe the steam treatment of soils. Either 'partial sterilization' or 'pasteurization' would be more correct as a temperature of 100 °C is not sufficient to kill the spores of bacteria and render the media sterile. A temperature of 125 °C is required to kill all of the organisms present in soils. In the USA, where steam–air mixtures are more widely used to achieve temperatures below 100 °C the term 'pasteurization' is commonly used.

12.1 THERMAL DEATHPOINTS

The temperature required to kill an organism is known as its thermal deathpoint. Usually a precise temperature cannot be stated, as this largely depends upon three factors: the form in which the pathogen is present (resting bodies and eggs are slightly more resistant to heat than the active stages); whether the heat being applied is moist or dry (moist heat is more effective than dry heat); and, the most important factor of all, the thermal deathpoint is closely related to the duration of the heating period. Commonly accepted values of thermal deathpoints are: 49 °C (120 °F) maintained for 10 minutes for nematodes and eelworms, and 50 °C (122 °F) for most of the fungi, although some wilt fungi require 82 °C (180 °F) for 10 minutes. Insects and slugs are killed at 66 °C (150 °F) and most weed seeds are killed at 77 °C (170 °F) maintained for 10 minutes; a few seeds, however, are known to survive at 100 °C (212 °F). Clover and *Oxalis* are examples of seeds that are difficult to kill by heat pasteurization. A temperature of 100 °C (212 °F) for 10 minutes is necessary to inhibit germination of *Oxalis*, but seeds can survive 82 °C (180 °F) for 10 minutes. Although tomato mosaic virus (TMV) present in root tissue is inactivated at 85 °C (185 °F) for 10 minutes, in dry debris it can survive a dry heat of 100 °C (212 °F) for 20 minutes. Sheard (1940) determined the thermal deathpoints of a number of fungi and eelworms under both laboratory and practical nursery conditions. He found that whereas pathogens growing in pure cultures in the laboratory were killed at 60 °C (140 °F) for 10 minutes, under typical nursery conditions, a temperature of 70 °C (158 °F) for 10 minutes was required to obtain a complete kill. This apparent increase in the thermal deathpoint under practical conditions was attributed to the difficulty of ensuring that *all* the soil had reached a temperature of 60 °C (140 °F); small pockets of soil which were slow to heat up allowed some survival of the pathogens. For this reason the higher temperature, which offers a safety margin, was recommended for nursery conditions.

The relationship of the thermal deathpoint to the duration of the heating period is of considerable practical significance to growers. By heating the soil to a lower temperature for a longer period, some of the chemical and biological reactions which occur at high soil temperatures, and which can be detrimental to plant growth, are thereby avoided or reduced. This relationship is shown in Figure 12.1 for the chrysanthemum

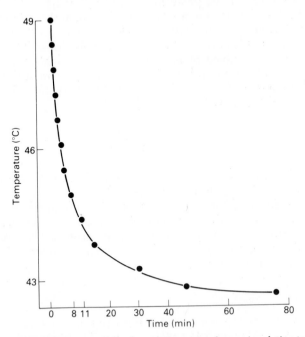

Figure 12.1 Thermal deathpoint of the chrysanthemum eelworm in relation to the time the temperature is maintained.

eelworm (*Aphelenchoides ritzema-bosi*). Although the actual values given are specific for this particular organism, the curve is typical of that found for a number of pathogens and shows how a long heating period at a low temperature can be used in place of high temperatures for shorter periods. The temperature × time relationship can be expressed mathematically as an inverse square root curve:

$$(Y - c)\sqrt{X} = K$$

where Y = the lethal temperature, X = the duration of the heating period, and c and K are constants.

Sometimes a pink or white fungus will be found growing on the surface of a heap of fresh potting soil a few days after it has been steam treated. These non-pathogenic fungi, *Pyronema confluens* (pink) and *P. glaucum* (white), are sometimes known as 'burnt earth' fungi as they can occur after prairie fires. *Peziza aurea*, which is shaped like an inverted mushroom with a dark hollow top and white on the underside, may occur when glasshouse soils are steamed and left uncultivated for a period.

12.2   METHODS OF HEAT PASTEURIZATION

The most commonly used methods of heat pasteurization are: steam, steam–air mixtures, flame pasteurizer, and electricity. In the past, two other methods have also been used, viz. baking and boiling water, but because of harmful side effects neither of these methods is recommended. Where a steam supply is available, e.g. from a glasshouse heating system, this is an obvious choice of method, especially if large quantities of pasteurized soil are required. When there is no steam supply or when only small quantities of sterilized soil are required, flame pasteurizers and electric sterilizers are convenient alternatives.

12.3   STEAM

The main advantages of using steam for soil pasteurization are: (a) it is a concentrated form of heat and is easily conveyed from the boiler to the sterilizing area, and (b) it gives control over the final soil temperature. With this form of heating, the temperature cannot rise above 100 °C (212 °F) and provided that the soil has been correctly prepared and the steam is 'dry', the treated soil can be used immediately the heating process has finished and the soil has cooled.

12.3.1   Steam generation

Steam produced from water boiling at atmospheric pressure has a temperature of 100 °C (212 °F) with 419 kJ kg^{-1} (180 BTUs lb^{-1}) sensible heat and 2257 kJ kg^{-1} (970 BTUs lb^{-1}) of latent heat. It is the latent heat in the steam which is available for sterilizing the soil. When steam is generated under pressure, i.e. in a boiler, the temperature at which the water boils is increased, thereby producing steam at a higher temperature and with a slightly increased amount of sensible heat; there is also a slight drop in the amount of latent heat. For example, steam at a pressure of 4 bars or 400 kPa has a temperature of 152 °C, with 640.7 kJ kg^{-1} of sensible heat and 2108.1 kJ kg^{-1} of latent heat. The volume occupied by 1 kg of steam also decreases with pressure, reducing from 1.673 m^3 at atmospheric pressure to 0.374 m^3 at 400 kPa^1. 1 lb of steam at 60 lb m^{-2} has a volume of 5.8 ft^3 whereas at atmospheric pressure the volume is 26.8 ft^3. The principal advantage to the grower of using high pressure steam is the decrease in the size of steam pipe required to link the boiler to the sterilizing area. When the steam is injected into the soil, it returns to practically atmospheric pressure and hence the soil temperature must be 100 °C (212 °F); there will be no difference in soil temperature resulting from the pressure at which the steam is used.

A secondary advantage of generating high pressure steam is the drying effect, or the reduction in the amount of free water present in the steam, which occurs as the pressure is reduced; this is technically known as the 'wire-drawing' effect. Steam generated in the manner described is not superheated and is known as 'saturated steam'. If it does not contain any free water droplets it is known as 'dry saturated steam'. In practice, however, boilers do not produce absolutely dry steam; also there is inevitably some loss of heat with condensation occurring in the steam mains. Depending upon the type of boiler, the length of the steam main and the thermal efficiency of the insulation, the steam which is used to heat the soil may contain up to 10% by weight of water. The additional heat present in the high pressure steam is used up by drying the steam or evaporating some of this free water as the steam pressure is reduced from 400 kPa to atmospheric or zero pressure. In practice, it is very important that the steam used for pasteurizing is dry. Wet steam can create puddled or waterlogged patches in the soil which result in either inefficient pasteurization due to lack of steam penetration, or soil which is too wet to handle. To ensure that the steam is dry, a steam separator should be used to remove any free water (§ 12.3.3).

12.3.2   Heat requirement of soils

The main factors determining the amount of steam required to pasteurize a given volume of soil are its heat capacity, and the thermal efficiency of the heating process.

The heat capacity of a soil comprises the dry matter, which has a specific heat of 0.837 J g^{-1} °C^{-1} and the soil moisture, which has a specific heat of 4.186 J g^{-1} °C^{-1}, thus to heat an oven-dry soil would require only one fifth of the amount of heat required to heat an equal weight of water. In practice, it is found that soils which are very dry (dust dry) can be difficult to steam because of the slow rate at which the steam is condensed. Wet soils, having a high heat requirement, are also difficult to steam satisfactorily because of the restricted spread of steam, 'cold spots' and puddling of the soil develop leading to the blowing of uncondensed steam through the surface. In practice, wet soils usually create more problems than dry soils. A light sandy soil in a suitable condition for steaming, having 12% by weight of water, has a heat requirement of 1744 kJ m^{-3} °C^{-1} (26 BTUs ft^{-3} °F^{-1}), whereas a heavy clay soil, with 28% moisture, has a heat requirement of 2146 kJ m^{-3} °C^{-1} (32 BTUs ft^{-3} °F^{-1}) (Morris 1954). The theoretical steam requirement per unit volume of soil will, therefore, be the product of the temperature rise times the heat capacity. For practical purposes we may say that an average theoretical heat requirement for potting soils would be 80 kg of steam per m^3 of soil (135 lb per yd^{-3}).

12.3.3 Thermal efficiency

The thermal efficiency of the heating process depends upon:

(a) the physical state and moisture content of the soil,
(b) the quality or dryness of the steam,
(c) the design of the equipment.

When steam is introduced into the soil, usually by means of a perforated pipe or grid buried in the soil, it immediately condenses on to the cold soil adjacent to the steam orifice and releases its latent heat, thereby raising the soil temperature to 100 °C (212 °F). As more steam is injected into the soil, it travels through the pore spaces in the already heated soil until it reaches cold soil where it condenses. In this way a very narrow heat front, only about 2.5 cm (1 in) thick, spreads through the soil. Irrespective of the steam pressure in the steaming grid, the pressure of the steam in the soil will not exceed 3.45 kPa (0.5 lb in^{-2}) and it will often be only half this value. For practical purposes, the steam can be considered as being at atmospheric pressure and hence at a temperature of 100 °C (212 °F). It therefore follows that, when steam is in contact with soil, the temperature of the heated soil must be 100 °C (212 °F). Whenever a temperature of 82 °C (180 °F) is quoted in connection with steam pasteurization this can apply only to very small areas where, because of a poor physical condition, steam penetration has not occurred and heating has been by conduction. Under good pasteurizing conditions, heating by conduction is minimal and the bulk of the soil will be at 100 °C (212 °F). If the free pore spaces in the soil are reduced either by compaction, lack of aggregate stability or by water, and the steam is unable to spread freely, the steam pressure in the soil builds up to the point where it blows up through the surface and escapes; for this reason a dry soil in good tilth is necessary for efficient pasteurization. Localized puddling of the soil and steam loss by 'blowing' can be caused by free water introduced into the soil by 'wet' steam. To ensure that the steam is 'dry', a steam separator must be installed at the end of the steam line. The separator operates on the principle of creating a rotary flow of the steam, which allows any water present to be separated by centrifugal force; this falls to the bottom of the separator, where it is released by a thermostatically controlled steam trap.

Assuming the above requirements regarding soil condition and steam quality have been met, the amount of steam required under practical conditions will range from 88 to 112 kg m^{-3} of soil (Bunt 1954a), i.e. the thermal efficiencies will be of the order of 90 to 70%. Long heating periods give lower thermal efficiencies because of the loss of steam from the sides of the sterilizing bin. Usually, rapid heating of the soil is desirable and on the basis of a ten-minute heating period and a steam requirement of 104 kg m^{-3} of soil, a steam flow rate of 624 kg h^{-1} will be required. Although 104 kg of steam is required to heat each metre of soil, the increase in the water content as a result of steaming will be equivalent to only approximately half the amount of steam used. This is partly because of some waste of steam during the heating process and partly because of the continued loss of water vapour from the soil after the steam has been turned off.

12.3.4 Treatment of boiler feed water

To prevent the corrosion of the steam mains and condensate return pipes, which is caused by oxygen and carbon dioxide dissolved in the water, it is customary to treat the boiler feed water with corrosion inhibitors, and two types of amines are commonly used for this purpose:

Filming amine (octadecylamine) is an inert chemical which volatilizes with the steam and condenses on the walls of the pipes, forming a film on the inner surfaces. Tests have shown that this material is not toxic to plants and it may be used safely as a corrosion inhibitor.

Neutralizing amine volatilizes and combines with the carbon dioxide in the steam. There have been several reported instances of phytotoxicity to chrysanthemums and poinsettias when grown in media treated with steam containing neutralizing amines. Horst et al. (1983) found that Chrysanthemum 'Indianapolis White' was susceptible to low concentrations of 2-diethylaminoethanol (DAE) in either the growing media or when the relative humidity was controlled by steam-generated mist. The symptoms included chlorosis, epinasty, stunting and necrosis. Other cultivars such as 'Bonnie Jean' and 'Mistletoe' were less susceptible. Phytotoxicity has also been reported from other neutralizing amines such as morpholine and cyclohexamine. Neutralizing amines should only be used in closed circuit heating systems, where no steam is ever used for sterilizing media. On glasshouse nurseries, it is always probable that steam will be wanted on some occasion to sterilize growing media and neutralizing amines are not therefore recommended as corrosion inhibitors for glasshouse boilers.

12.4   STEAM–AIR MIXTURES

There are several reasons why it is desirable when heat treating soil that a temperature of less than 100 °C (212 °F) should be used, the most important being:

(a) biological changes are reduced at the lower temperatures,
(b) changes in the organic and mineral form of plant nutrients following heating are reduced,
(c) less heat is required and therefore the cost is lower.

It has already been shown that the thermal deathpoint of soil-borne pathogens can be safely reduced below 100 °C by extending the length of the heating period (see Fig. 12.1). Biologists have long been aware that the natural biological control of diseases plays an important role, i.e. one organism is antagonistic to, and controls the numbers of another. When soils are heated to 100 °C, both the pathogens and their natural antagonists are eliminated. Therefore if pathogens reinfect a recently steamed soil they spread at a much higher rate because of lack of competition and cause a greater loss of plants than if the soil had not been treated. Baker (1967) has described the situation occurring in recently steamed soils as being a 'biological vacuum'. He contrasts the loss of seedlings grown in a soil which had been steam pasteurized and then inoculated with the damping-off fungus *Rhizoctonia solani*, with another treatment where *Myrothecium verrucaria* was introduced at the same time as *R. solani*. In the first treatment, the pathogen destroyed most of the seedlings whereas, in the latter treatment, very few of the seedlings were killed because of antagonism and the containment of the pathogen by the other fungus.

One way of countering the problem of reinfestation is to reduce the temperature to which the soil is heated and so allow the natural antagonists to survive the heat treatment. It has already been shown that, when using steam as the heating medium, a temperature of 100 °C in the bulk of the soil is inevitable; by introducing air into the steam supply, a steam–air mixture is created which has a lower temperature. On the occasion of the first recorded use of steam–air mixtures for the heat pasteurization of soil (Bunt 1954b), an industrial pattern steam injector operating on the *venturi* principle was used to entrain free air. The temperature of the mixture could be controlled by variable orifice steam- and air-nozzles. Since that time, Baker and his co-workers have made detailed studies of both biological and engineering aspects of steam–air mixtures, and this method of soil pasteurization is now well established in the USA and Australia.

12.4.1   *Steam and air requirements for various temperatures*

The flow rates of steam and air required to treat one cubic metre of a 1:1:1 soil:peat:perlite mix at temperatures from 60 °C (140 °F) to 100 °C (212 °F)

Table 12.1 Rates of steam and air flow required to treat 1 m^3 of medium in 30 minutes (50% heating efficiency assumed). (Adapted from Aldrich *et al.* 1972.)

Mixture temperature (°C)	Steam (kg min^{-1})	Air (l min^{-1})
60.0	1.833	10116
65.5	2.070	8195
71.1	2.306	6534
76.6	2.542	4983
82.2	2.838	3618
87.7	3.253	2362
100.0	3.843	0

are given in Table 12.1. Equipment suitable for treating media with steam–air mixtures, using either continuous flow or batch processes, has been described by Aldrich (1976).

To avoid large cumbersome steam injectors, it is customary to use either air blowers or fans to supply the air for the mixture (which should have a temperature of 66–70 °C). It should be noted that, unlike pasteurization by pure steam where rapid heating and cooling is desirable, with the steam–air method, the temperature must be maintained for thirty minutes. The steam–air mixture can be injected into the soil either from a buried grid, which allows the heat-front to rise up through the soil, or under a plastic sheet held over the surface of the soil so that the heat-front travels downwards from the soil surface. The latter method is preferred as 'blow-out' problems caused by lack of lateral spread of the mixture are avoided.

Using this system of heat pasteurization Baker (1967) has shown that the number of plant deaths following the reintroduction of *R. solani* into heated soil were significantly reduced when a temperature of 70 °C (160 °F) had been used, and no losses occured at all when the temperature used was 60 °C (140 °F). This effect was attributed to the survival of the natural antagonists at the reduced soil temperatures.

Changes in the forms of nitrogen and amounts of exchangeable manganese produced by this method are less than those occuring from the use of pure steam; this is discussed more fully in a later section on the chemistry of pasteurized soils.

12.5 FLAME PASTEURIZER

Another method of heat treatment of soil is by the flame pasteurizer. This consists of a revolving steel cylinder mounted at a slight angle to the horizontal; the cylinder is open at both ends and rotates at about 50 r.p.m.

(Fig. 12.2). Heat is supplied by a blow-torch, burning either paraffin, a light oil or gas. The latter fuel is recommended as a temporary blockage in the fuel line can cause a 'flame-out', resulting in unburnt paraffin or oil discharging into the soil and causing phytotoxicity. The mode of pasteurization is primarily by steam generated from the water present in the soil rather than by direct heat. Provided that the soil is sufficiently moist and that the equipment is used correctly, the soil temperature will not rise above 100 °C (212 °F). It can be demonstrated that no significant amount of burning of the soil occurs, by including small pieces of paper or dry straw with the moist soil. If the equipment is being operated correctly, the paper will not be charred or burnt.

In operation, moist loose soil is shovelled into the upper end of the cylinder and then falls against the hot sides as the cylinder revolves. Soil is in the cylinder for only 20–30 seconds and the temperature can be controlled by varying the size of the flame, the angle of the cylinder and the amount of water in the soil. If the soil is too wet, however, it will 'ball-up' and will not mix uniformly with the other ingredients. To achieve satisfactory results, the pasteurizer should be operated to give a soil temperature of 79 °–88 °C (175 °–190 °F); a temperature of 70 °C (160 °F) has not proved satisfactory because of the very short time the soil is in the pasteurizer. This method of pasteurization has the advantage of being a 'flow' rather than a 'batch' process, with consequently a high potential output. Pasteurizers are available with outputs of up to 6 m^3 hr^{-1}; the fuel requirement is about 6 l m^{-3} of soil treated.

Figure 12.2 Mobile flame pasteurizer. Rotating screen at end of drum (covered with a shield) sieves media leaving the pasteurizer.

12.6   ELECTRICAL PASTEURIZERS

Electricity can be used in two distinct ways to heat pasteurize soil, either:

(a) indirectly, by means of heating elements of the immersion heater type,
(b) directly, by passing an electric current between two electrodes buried in the soil.

12.6.1   *Immersion heater type*

Heat can be generated by passing an electrical current through a wire having a high electrical resistance. If the heating element is mechanically protected and also electrically insulated, it can be used to heat pasteurize soil in much the same way as an immersion heater is used in a domestic kettle. The only essential difference is that whereas fluids, such as water, can circulate in relation to the heater, thereby producing convection currents, soils are not fluid and the heating is almost entirely by conduction. A small amount of heating does occur by steam generated from the moist soil which is in direct contact with the heaters.

The pasteurizer is usually in the form of a heat-insulated box mounted on legs, the base being hinged and held in the closed position by quick-release clamps; this allows rapid removal of the soil after treatment. The heating elements consist of sheathed resistance wires embedded into large plates. In addition to giving mechanical protection to the elements, the plates also provide a large heating surface.

With this type of pasteurizer, heating is by conduction, and dry soils have a relatively low rate of heat conduction. When the particles of soil are joined by a film of water, the rate of conduction is appreciably increased but, as the heat capacity of the soil is also increased, a compromise in the soil water content must be reached. Kersten (1949), working with a silty-clay loam (27% clay), found that, at a moisture content of 12.3%, the thermal conductivity was $0.0084 \, \mathrm{J \, s^{-1} \, cm^{-1} \, {}^\circ C^{-1}}$ and the heat capacity was $1.825 \, \mathrm{J \, cm^{-3} \, {}^\circ C^{-1}}$. When the moisture content was increased to 18% the thermal conductivity was increased by 25% and the heat capacity by 18%. However, as the soil was made progressively wetter the thermal conductivity did not rise as rapidly as the heat capacity. Therefore, soil to be pasteurized by this method should not therefore be very wet.

The surface temperature of the heaters rises to about 300 °C (572 °F), resulting in a thin layer of overheated soil, while the bulk of the soil reaches only 70 °C (158 °F). In practice, this small volume of overheated soil does not appear to cause any harmful effects to seedling growth. The reason for this is not clear. Dilution of the small volume of overheated soil with the remainder of the soil which has not been overheated has been one suggestion; it also seems probable that, because of the very high temperatures attained in the soil, toxins are not produced at the same rate as when soils are heated to intermediate temperatures (§ 12.8.1). One advantage of this type of pasteurizer is that the electrical loading is constant throughout the heating period; also the method is quite safe because the soil is never electrically 'live'.

Pasteurizers of this type are available in two sizes, one which is suitable for commercial use has a capacity of 0.2 m^3 ($\frac{1}{4}$ yd^3) of soil and an electrical loading of 9 kW, a smaller version suitable for the amateur has a capacity of 0.04 m^3 (1.3 ft^3, 1 bushel) and an electrical loading of 1.5 kW.

12.6.2 Electrode pasteurizers

An alternative method of electrical soil pasteurization is to pass an electrical current between two electrodes buried in the soil; the water acts as an electrical conductor and a resistance heater. With this method, the soil is electrically 'live' and adequate safety measures, such as an interlocking switch on the lid of the pasteurizer, must be employed to protect the user. Unlike the immersion heater type of pasteurizer, the electrical loading is not constant during the heating process. At the start there is a high electrical resistance and the current is low. As the heating progresses, the resistance drops and the current rises. With some designs of pasteurizer, the flow of current just before the heating is finished can be up to eight times as high as at the start; under some conditions of electrical supply this can be a considerable disadvantage. The electrical resistance of a particular soil is directly proportional to the distance apart of the electrodes and inversely proportional to their area. Normally, the electrodes are 25–30 cm (10–12 in) apart. With some pasteurizers it is possible to halve the current being used by switching alternate electrodes out of circuit after the soil has started to heat.

Another factor crucial to the success of the operation is the electrical conductivity of the soil. This is governed by the amount of moisture and salts it contains and also by its density. Dry, loose soil is a very poor electrical conductor, whereas a soil that is moist and contains some mineral salts is a good electrical conductor. During pasteurization, the soil can contract away from the electrodes, and firm packing when filling is necessary to maintain electrical contact. Soil to be pasteurized by this method should therefore be moist, contain about 35% water and be firmly packed. If the soil has a low mineral salt content, a solution of magnesium sulphate (6 g l^{-1} or 1 oz gal^{-1}), applied with a watering-can as the pasteurizer is being filled, will improve the conductivity and will not cause any harm to seedling growth. Heating ceases when the soil adjacent to the electrodes dries out, thereby breaking the flow of electricity; at this point, the temperature of the bulk of the soil will be about 77 °C (170 °F). Pasteurizers of this type are available with a capacity of 45 l (1$\frac{1}{4}$ bushels), operating on single phase 240 V a.c. system with a maximum loading of 25 A. A larger three-phase model with just over 380 l ($\frac{1}{2}$ yd^3) capacity has a maximum loading of 40 A per phase. With both the immersion heater and electrode type of pasteurizers, the electrical consumption is just over 1 kW per 28 litres (1 ft^{-3}) of soil, but this figure is very much dependent upon the design of the equipment and the condition of the soil.

In recent years, the increasing use of loamless mixes which do not normally require pasteurization together with the increased number of nurseries having a steam supply available, have resulted in a decline in the use of electrical soil pasteurizers.

12.7   OTHER METHODS

Three other methods of pasteurizing soil can be mentioned: baking and boiling water (largely for historical interest), and the more recent developments with radiation treatments.

Baking was widely used in the early days when steam and electrical equipment were not generally available. Moist soil was placed on a steel plate over a fire; at the commencement of heating, the soil was in fact being steamed by steam generated from the water within the soil. During the later stages of heating, when the water had been evaporated, there was no control over the soil temperature, and often very high temperatures occurred. In some soils, marked toxicities to seedling growth resulted and it was necessary after pasteurization to leave the soil to 'recover' before it could safely be used.

Boiling water has also been used for heat pasteurizing soils, but its low heat content and the puddled state of the soil after treatment did not commend its use.

Rattink (1982) used gamma radiation at low dosages, ca 250 krad, to kill *Fusarium, Sclerotium, Phytophthora* and weed seeds without any phytotoxicity. Van Wambeke *et al.* (1983) treated shallow layers of soil and rockwool with microwave radiation, i.e. 2450 ± 20 MHz, to inhibit the growth of *Rhizoctonia, Fusarium, Verticillium, Phyrenochaeta, Trichoderma* and nematodes. However, these methods of soil pasteurization are confined to the laboratory at present and are not used to treat the volume of media required on a nursery scale.

12.8   CHEMISTRY OF HEAT PASTEURIZATION

It has long been known that substances toxic to plants can be produced when soils are heated. The extent to which seed germination and seedling growth is retarded in pasteurized soil can be very variable. In some cases it is so small that it passes unnoticed, especially as the 'sterilizing check' to plants is usually only temporary and this is quickly followed by a period of faster growth, which soon produces larger plants than those grown in untreated soil. In other cases, the check can be so great that it leaves a lasting effect on the growth of such plants as *Alyssum* and *Antirrhinum*; in extreme cases it may even result in their death.

It was mistakenly thought at one time that only one specific toxin was produced when soils are heated and attempts at its identification lead to apparent contradictory results. Although the full extent of the chemical and biological changes following heat pasteurization are still not known, it is possible to group the probable causes of toxicity and to suggest remedial action.

259

12.8.1   Nitrogenous compounds

Early workers (Pickering 1908, Johnson 1919) attributed the toxicity to an increase in the soluble organic matter and ammonia production. When soils are heated to 100 °C, the nitrifying bacteria are killed and even if re-inoculation is made immediately after pasteurization, they do not establish and multiply for some time. The spore-forming ammonifiers, however, are not suppressed and following heat pasteurization there is usually a build up of ammonium before nitrates are once again formed in significant amounts. Johnson examined the effect of different temperatures on the production of toxic substances. In soil heated to 50 °C, there was no delay or inhibition in the germination of lettuce seed by comparison with seed sown in the untreated soil. Between 100 °C and 250 °C, however, there was a considerable delay in germination; but when the sterilizing temperature was increased to 350 °C, germination improved again and showed only a slight delay by comparison with the control treatment. The effect of the temperature of the heat treatment on seed germination was closely related to the amount of ammonium produced; this rose from 2.5 mg per 100 g in the unheated soil to 13.9 mg in the soil heated to 250 °C, and then declined to 7.3 mg in the soil heated to 350 °C. He was also able to show that when working at a temperature of 110 °C plant growth (weight) decreased as the heating period extended beyond ten minutes. However, there was not always consistency between results obtained from different soils; tomato plants grown in peat heated to 110 °C showed a decrease in growth of 200% by comparison with plants grown in unheated peat, whereas with a silt loam, heating to this temperature gave a 254% increase in growth. It was suggested that ammonia was the toxic agent. This work was extended by Lawrence at the John Innes Horticultural Institution in the 1930s, and resulted in a better understanding of the practical requirements for successful soil pasteurization.

Recent work has shown conclusively that high concentrations of ammonium (Maynard & Barker 1969, Nelson & Hsiek 1971) and of ammonia (Bennett & Adams 1970) in soils can be toxic to plants. Below pH 6.0 there will be very little free ammonia; it will mostly be present in the ammonium (NH_4^+) form. As a result of heat pasteurization and the temporary blockage in the nitrogen cycle, ammonium produced by the biological decomposition of the natural organic matter present in the soil and from such nitrogenous fertilizers as hoof and horn, urea-formaldehyde, etc., can result in a temporary increase in the pH of the media to the neutral point and above. Significant quantities of free ammonia can then be present.

Other forms of nitrogen can also be toxic to plants. Usually ammonium is converted to nitrites, which are then rapidly converted to nitrate nitrogen, and only very small or trace amounts of nitrite nitrogen will be found in normal soils. *Nitrobacter*, which converts nitrite to nitrate, is killed by steam pasteurization and is also inhibited by the presence of high levels of ammonia. Consequently nitrites may accumulate to toxic levels for a short period in steamed soils. Court *et al.* (1962) show that, in addition to the risk of free ammonia toxicity occurring when urea was used as a nitrogen source, nitrites could also cause toxicity. Using a bioassay technique, they found that when the nitrite nitrogen in the soil reached 110 p.p.m. the growth of young maize seedlings was reduced to 30% of that of the control plants. Sonneveld (1979) found that steam pasteurization not only increased the amount of exchangeable ammonium nitrogen, it also reduced the amount of nitrate nitrogen and slightly increased the nitrite nitrogen. The degree of nitrate decomposition depended upon the duration of the steaming treatment and the temperature reached. Although ammonium, free ammonia and nitrite are the forms of nitrogen which are generally believed to be of most consequence following heat treatment, several workers have cited toxicities arising from other nitrogenous compounds.

12.8.2 Soluble organic compounds

Some soils release toxic amounts of soluble organic matter when they are heat treated. Walker & Thompson (1949) found that after steam pasteurization a peat soil gave an increase of 650% in the amount of soluble organic matter present, whereas another soil with a low organic matter content (loss on ignition 11.4%) gave only a 25% increase in the soluble organic matter following steaming. Schreiner & Shorey (1909) isolated dihydroxystearic acid from organic soils and found it injurious at all concentrations and lethal at 100 p.p.m.

12.8.3 Manganese

In addition to the increase in the amount of soluble organic matter and the various forms of nitrogen following the heat treatment of soils, the solubility of several minerals including calcium, zinc, potassium, copper, aluminium and manganese is also increased; of these, manganese is usually the most important from the phytotoxicity point of view. Early workers noted the large increase in the soluble and exchangeable forms of manganese following steam pasteurization.

Manganese can be present in soils either as relatively inert or as easily reducible compounds. In the latter group the tri- and tetravalent forms are readily reduced to the more soluble bivalent or exhangeable form by steaming, and heat treated soils usually have much higher amounts of readily available manganese. After steaming, the manganese is converted again to manganese oxides by bacteria, but this process may occur rather slowly, especially if the pH is low. In some soils it has taken 200 days before the exchangeable manganese has dropped to the level present before steaming. Sonneveld (1979) followed the changes in the levels of exchangeable manganese in a steamed soil and in the same unsteamed soil to which manganese sulphate had been added to give similar exchangeable manganese levels of about 140 p.p.m. In the unsteamed soil the exchangeable manganese dropped rapidly to less than 50 p.p.m. in

15 days. In the steamed soil, however, the same level was only reached after 120 days.

Davies (1957) found that steam pasteurization of a brickearth soil (pH 4.26) resulted in an immediate increase in the exchangeable form of manganese, and 11 days after steaming the manganese level had risen from 106 to 1444 p.p.m. The effect of soil pH on manganese release was studied by applying calcium carbonate at varying intervals before pasteurization. It was found that the longer the interval between liming and pasteurizing, the smaller was the increase in the amounts of water-soluble and exchangeable manganese produced. Increasing the soil pH from 5.1 to 7.0 reduced the amount of manganese in the tissue of young tomato plants from 4900 p.p.m. to 1300 p.p.m., and the addition of superphosphate to the soil reduced the manganese level still further to 800 p.p.m.

Messing (1965) found that the pasteurization of soil at pH 5.9 produced a much smaller release of manganese than when the soil was at pH 5.3. Furthermore, below pH 5.3, the amount of extractable manganese, i.e. the water-soluble plus the exchangeable manganese, increased during the period of the experiment, whereas above pH 5.9 the manganese levels decreased during the experiment. The effect on the uptake of manganese by plants of adding superphosphate to the steamed soil, was found to be dependent upon the pH of the soil. If the pH was high, then adding superphosphate invariably decreased the pH of the soil and increased the amount of manganese absorbed, but when the pH was low, applying superphosphate reduced the amount of manganese taken up by the plants. Where manganese toxicity is expected to occur following heat treatment, application of a solution of 0.6 g l^{-1} (0.1 oz gal^{-1}) of mono-ammonium phosphate (= 150 p.p.m. phosphorus) will often give beneficial results.

For advisory purposes, Harrod (1971) has used the amount and forms of manganese present in the soil, either before or after pasteurization to estimate the risk of manganese toxicity to plants (Table 12.2). It is assumed that some of the easily reducible manganese will be converted into the exchangeable form as a result of steaming. Most workers are agreed that there is little risk of manganese toxicity resulting from steam pasteurization provided that the soil pH is 6.0–6.5 before steaming.

Table 12.2 Manganese toxicity risks.

Risk	Manganese levels (p.p.m.)	
	before steaming; exchangeable plus easily reducible forms	after steaming; exchangeable form
safe	100	<75
doubtful	100–250	75–100
dangerous	250	>100

12.8.4   Aluminium

During the course of his investigations into the manganese toxicity of lettuce grown in steam pasteurized soils, Messing found that aluminium could also be a related factor. Increases in the amount of extractable aluminium, after steaming, ranged from 15 to 100% of that found in the unsteamed soil. Those measures already recommended to control manganese toxicity, i.e. liming to pH 6–6.5, and applying superphosphate to the soil, will also control aluminium toxicity.

12.8.5   Effects of pasteurizing temperature

Several workers have found that when steam–air mixtures are used to pasteurize the soil at temperatures below 100 °C, i.e. the temperature obtained using steam, the degree of chemical changes in the soil is less. For example, manganese levels in soils 3 days after being heated to different temperatures (Dawson et al. 1965) are given in Table 12.3. At 10 days after the soil was heated to 82 °C there was only 5 p.p.m. of manganese present, whereas in soil heated to 100 °C the corresponding value was 50 p.p.m.

White (1971) showed that there was less ammonium nitrogen produced in soils that were heated to 60 ° and 71 °C than at 100 °C. There was also less nitrite nitrogen: 42 days after heating to 100 °C, 71 ° and 60 °C the nitrite nitrogen values were 22, 14 and 13 p.p.m. respectively. The beneficial effects of using a lower temperature to pasteurize soil in which carnations were grown varied with the nitrogen source. With urea, treatment at 60 °C gave 23% more fresh weight than heating to 100 °C, with a predominately nitrate nitrogen source, however, there was only 5% more growth in the lower temperature soil treatment. There was also an interaction between the nitrogen source, the heating temperature and the amount of manganese in the leaf tissue. At all pasteurizing temperatures nitrate nitrogen gave higher manganese levels in the tissue than did urea, but heating the soil to 60 °C rather than 100 °C reduced the manganese level more with urea than it did with nitrate nitrogen.

The weight of lettuces grown in soil treated at 70 °C was about 20% greater than those grown in soil heated to 100 °C, quality was also improved and tipburn reduced (Sonneveld & Voogt 1973).

Table 12.3   Effect of the pasteurizing temperature on the release of manganese.

Soil temperature	Water-soluble plus extractable manganese (p.p.m.)
control (unheated)	8
60 °C (140 °F)	11
71 °C (160 °F)	15
82 °C (180 °F)	27
88 °C (190 °F)	38
93 °C (200 °F)	58
100 °C (212 °F)	83

12.8.6 Pasteurization and controlled-release fertilizers

The release rate from most controlled- or slow-release fertilizers is significantly increased when the media temperature is raised, thereby reducing the controlled-release effect and creating either salinity problems or nitrogen toxicities. Where the release action is primarily controlled by the rate of solubility, e.g. MagAmp, the effect of steaming the mix with the fertilizer already incorporated will be much less than with other types of fertilizers. However, it is advisable not to add any controlled-release fertilizers to media before pasteurizing, it is much safer if they are added afterwards.

12.9   RULES FOR HEAT PASTEURIZATION

Observance of the following rules when treating soils will ensure the maximum benefits with the minimum risks:

(a) Before a new and unknown soil is steamed and used on a large scale, first determine its response to heat pasteurization and the risk of nitrogen and manganese toxicity by growing a trial amount of seedlings, such as antirrhinum.
(b) If steam is used, ensure that both the soil and the steam are 'dry'. Excessively wet soil and wet steam cause inefficient heating and leave the soil in a poor physical condition. Heat the soil rapidly (within 10 minutes), and, after allowing a further 10 minutes with the steam turned off, remove the soil and cool it quickly.
(c) Ensure that the soil pH is between 6.0 and 6.5, avoid soils with a high organic matter content and do not add large amounts of organic nitrogen in the base fertilizer. Early commencement of liquid feeding is better and safer than trying to give too much nitrogen in the base fertilizer.
(d) If the sterilized soil or the made up medium has to be stored before use, *it must be kept at a low temperature and dry* to minimize biological changes which occur in the different forms of nitrogen.
(e) Low-temperature steam–air mixtures give less risk of toxicity than high temperature steam, *but the length of the heating period must be adjusted to the temperature of the steam–air mixture: 71 °C (160 °F) for 30 minutes is recommended.*

Heat pasteurization offers the grower the means of ensuring a pest- and disease-free medium with which to commence the propagation of seedlings and the growing of young pot plants. Provided that the above rules are observed, pasteurization checks to plant growth will be minimal and the benefits will far exceed any disadvantages. The increasing popularity of mixes made from materials which do not require heat treatment probably means that the peak of interest in soil pasteurization for pot plants has passed. Specific pest and disease problems will nevertheless continue to occur and it may well be necessary to resort to the heat pasteurization of these materials at times.

CHAPTER THIRTEEN

Chemical sterilization

By comparison with mineral soils, many of the materials used to make loamless mixes can be considered already sterile as far as plant pathogens are concerned. Some materials, such as foam plastics, perlite and vermiculite, have been subjected to very high temperatures during their manufacture and have virtually been sterilized already. Although many peats are free of pathogens, some samples may contain *Rhizoctonia* and *Pythium*, and there is a slight risk that bark samples may be contaminated with *Phytophthora* spp. However, the incidence of fungi in potting mixes made from peat or bark is low and it is common practice in Britain not to treat these with heat or chemicals except when growing certain crops which are particularly susceptible to soil-borne diseases. For example, peat used for cucumber propagation is often heat pasteurized or treated chemically against *Pythium* attack; *Antirrhinum* is another plant which is susceptible to *Pythium*.

Whenever there is the risk of pest or disease infestation in the bulky materials used in potting mixes and means of heat pasteurization are not available, chemical treatment can be considered as an alternative. Chemicals have the advantage of a low capital cost by comparison with steam and other forms of heat treatment. Their main disadvantage is that they often leave toxic residues, and treated material must be left for a period before it can safely be used.

13.1   SOIL FUMIGANTS

In addition to the materials traditionally used for the chemical sterilization of potting soil, e.g. formaldehyde, there are now several other chemicals which may be applied, as volatile liquids or powders, to sterilize materials by fumigant action.

13.1.1   *Chloropicrin*

Chloropicrin or tear gas is effective against most fungi and insects, but is not as effective as methyl bromide against nematodes. It is a heavy, almost colourless liquid which volatilizes when injected into the soil. *As the vapour is very pungent and can quickly cause nausea and tears, a respirator must be worn.* It is injected into the soil at the rate of 3–5 ml per 30 l and the soil must then remain covered with a gasproof cover for 24 hours. The soil must be in good physical condition to allow uniform penetration and must have a temperature of not less than 16 °C (60 °F). Chloropicrin fumes are toxic to living plants and all traces of the gas must be removed from the soil before it is used for making composts. After treatment, an interval of 2–4 weeks is required before the soil can be safely used. The higher the soil temperature and the smaller the heap, the shorter is the period required to remove the fumes. The area chosen for treating the soil should also be well away from growing plants. In the USA, chloropicrin is available as 'Larvicide' and 'Chlor-O-Pic'.

13.1.2   Formaldehyde

Commercial formaldehyde is an approximately 40% solution, known as formalin. It is diluted at 1 in 50 and applied at 150–300 l m^{-3} to saturate layers of the soil (150 mm thick) as the heap is built. The heap is covered with plastic sheeting for at least two days, then uncovered and turned. The soil must be left for 4–6 weeks before use, depending upon temperature, soil texture and number of turnings.

13.1.3   Methyl bromide

This material is effective against most fungi, insects, nematodes and weed seeds. Experience in the USA suggests, however, that it does not give adequate control of *Verticillium*. It is an almost odourless and colourless gas at normal temperature and is *highly toxic to humans;* usually 2% of chloropicrin (tear gas) is added as a warning agent. In Britain, the material can be applied only by approved contractors and *it is essential that the necessary safety precautions are fully observed and implemented.*

The soil to be treated is enclosed in a polythene sheet and the gas, liquefied under pressure, applied at the rate of 0.7 kg m^{-3} of soil (4 lb per 100 ft^{-3}) through a vaporizer. The soil should be moist and the temperature must not be less than 10 °C (50 °F). After treatment, the soil must remain sealed in the sheet for 4–5 days and then be freely ventilated for 4–10 days before it is used. The fumes are slightly toxic to some plants but the greatest risk is caused by the formation of inorganic bromides in the soil. Plants such as *Antirrhinum*, carnation and *Salvia* have been found to be particularly susceptible to bromide toxicity. The toxic residues formed in the soil can be removed by leaching but very heavy rates of water application are required. In the USA a mixture of methyl bromide with 33% chloropicrin, known as 'Dowfume MC-33', is available. This is more effective against *Verticillium*.

The dangers in using methyl bromide, requiring the employment of specialist operators, the problem of toxic residues and the difficulty in their removal from organic materials, all combine to discourage its use on materials to be used for making potting mixes. Methyl bromide is, however, widely used to treat glasshouse soils.

13.1.4 Na N-methyldithiocarbamate (metham-sodium)

This material has a general fungicidal, insecticidal and nematicidal action. Small quantities of potting soil can be treated by making up a 1% solution of metham-sodium (2 pints of 33% metham-sodium in 25 gal) and applying this to the moist soil in 15 cm (6 in) layers at the rate of $30\,l\,m^{-3}$ ($5\,gal\,yd^{-3}$). The metham-sodium breaks down in the soil to form methyl isothiocyanate. In addition to the liquid formulation, which is available as 'Sistan' or 'Vapam', a material having similar action in the soil, and known in the UK as dazomet, is available in prilled form. The prills are mixed uniformly with the moist soil at the rate of $220\,g\,m^{-3}$ ($6\,oz\,yd^{-3}$). After treatment, the soil is covered for three weeks and then turned three times at 14-day intervals. The vapour is very toxic to plants and it is essential that all traces of the methyl isothiocyanate have disappeared from the soil before it is used. This is readily tested by comparing the germination of cress seed in the treated and untreated soil.

In addition to the above-mentioned general purpose fumigants dichloro-propene-dichloropropane (DD) is an effective nematicide.

All of the above chemicals kill the nitrifying bacteria in the soil, producing an accumulation of ammonium nitrogen. Although their effect on the nitrogen cycle is not as great as that produced by heat sterilization, the same precautions regarding toxicities should be followed. Gasser & Peachey (1964) have reported that dazomet had a greater effect on retarding nitrification than metham-sodium. Methyl bromide was also found to increase the rate of mineralization of soil organic nitrogen more than the other sterilants.

13.2 SOIL FUNGICIDES

Because the low risk of serious fungal and insect infestations arising from materials used to make lightweight potting mixes, they are seldom pasteurized. A number of fungicides are however used for *prophylactic* or *preventative* control, they inhibit the spread of mycelium in the media. The fungicides are either added to the media during mixing or applied as a drench immediately after the seedlings have been pricked out; several of the newer compounds act systemically within the plant.

13.2.1 Benomyl

Benomyl ('Benlate') can be used as a soil drench at 1.1 kg per 1000 l (1 lb per 100 gal) for the control of *Fusarium*, *Rhizoctonia* and *Thielaviopsis* in *Poinsettia*.

13.2.2 Cheshunt compound

Although one of the early fungicides, this still provides a cheap and effective means of preventing *pythium* attack. It can be purchased ready-mixed, or prepared by mixing 57 g (2 oz) of finely ground copper sulphate with 312 g (11 oz) of ammonium carbonate and then stored for at least 24 hours in a tightly stoppered glass jar. A solution of this is prepared by mixing 28 g (1 oz) in a little hot water and making up to 9 l (2 gal).

13.2.3 Chlorothalonil

This is known as 'Daconil 2787', 'Bravo' and 'Repulse'. It has some suppressive action against *Rhizoctonia*.

13.2.4 Etridiazole or ethazol

Trade names for this fungicide are 'Truban', 'Terrazole' and 'Aaterra'. It is used to control *Pythium* and *Phytophthora* in pot and bedding plants. It is most effective when mixed into the media at 40 g m^{-3} (1 oz yd^{-3}) in peat mixes, and 75 g m^{-3} (2 oz yd^{-3}) in loam composts. For hardy nursery stock it is used at 180 g m^{-3} (4$\frac{1}{2}$ oz yd^{-3}). It can also be used as a drench at 19–62 g per 100 l (3–10 oz per 100 gal). Repeat applications of the drench may be given at 4–12 week intervals as required.
Note: apply the *lower* rates to soil-less mixes and rinse the foliage of seedlings with clear water.

13.2.5 Fosetyl-aluminium

This systemic fungicide is available as 'Aliette', it is either mixed into the media or applied as a soil drench for the control of *Phytophthora* stem and root rots of pot plants and hardy nursery stock. It is used at 241–482 g m^{-3} (6$\frac{1}{2}$–13 oz yd^{-3}).

13.2.6 Furalaxyl

This is a systemic fungicide available in Britain as 'Fongarid'. It can be incorporated into the potting media or applied as a soil drench for the control of *Pythium* and *Phytophthora* spp. which attack pot plants, bedding plants and hardy nursery stock. Either add 200–400 g m^{-3} (5$\frac{1}{2}$–11 oz yd^{-3}) mixed with dry sand as a carrier, or drench with 200–400 g in 50 l m^{-3} (5$\frac{1}{2}$–11 oz in 11 gal yd^{-3}).

13.2.7 Iprodione

Available in Britain as 'Rovral', it is particularly effective against *Rhizoctonia*, *Sclerotinia* and *Botrytis*. Apply as a drench, 40 g per 100 l (3 oz per 50 gal) using 5 l m^{-2}.

13.2.8 Metalaxyl

A systemic fungicide, sold in the USA as 'Subdue' for the control of *Pythium* and *Phytophthora* in pot and bedding plants. Apply as a drench at 4–15 g l^{-1} (0.5–2 oz per 100 US gal), for *Chrysanthemum* and geranium use the lower rate.

13.2.9 Propamocarb

Has a limited systemic activity; available in the USA as 'Banol' and 'Prevex', and in Britain as 'Filex' for the control of those *Phycomycete* fungi attacking seedlings, cuttings, bedding and pot plants. Drench at 1.5–$2.2\,g\,l^{-1}$ (20–30 oz gal^{-1}).

13.2.10 Quintozene

The active ingredient of this material is pentachloronitrobenzene (PCNB) and it is known as 'Botrilex' in Britain and 'Terraclor' in the USA. It is effective against *Rhizoctonia* and *Sclerotinia* and is mixed into the media at a rate of $890\,g\,m^{-3}$ ($1.5\,lb\,yd^{-3}$) two or three days before planting.

Sometimes mixtures of fungicides are made to give a wider spectrum of activity. For example, 'Terradactyl' is a mixture of etridiazole (or ethazol) and chlorothalonil. It is effective against *Pythium*, *Phytophthora* and *Rhizoctonia* when used at 100–$130\,g\,m^{-3}$ ($3\frac{1}{2}\,oz\,yd^{-3}$). 'Banrot' is a mixture of 15% etridiazole and 25% thiophanate-methyl, and is used to control *Pythium*, *Phytophthora*, *Rhizoctonia*, *Fusarium*, *Verticillium* and *Thielaviopsis*. It can be incorporated into the potting mix or applied as a drench; the dosage rate depends upon the plant species. 'Zyban' is a mixture of thiophanate-methyl and mancozeb and has a broad spectrum activity.

Caution The formulations of the various fungicides mentioned in this section and their percentage of active ingredients varies with the manufacturer. Because of this, precise directions on the strength of application cannot be given. *It is essential that the manufacturer's instructions are followed.* It is also advisable to test the materials first on a small scale. Where there is an acute disease problem and the reaction of the various plant species to the fungicides is unknown, it is preferable to use heat to sterilize the materials before making up the medium.

13.3 SOIL INSECTICIDES

Normally, the materials used in potting mixes are not likely to contain insect pests, and no treatment is required. If peat-based mixes have been stored for a period before they are used, there is the possibility of an infection of sciarid fly (fungus gnats) occurring. These insects are attracted by decomposing organic matter and are particularly troublesome in mixes containing organic sources of nitrogen, e.g. hoof and horn. As a preventative measure a wettable powder formulation of diazinon can be mixed into the media at $170\,g\,m^{-3}$ ($4\frac{1}{2}\,oz\,yd^{-3}$). This will control the fly for 4–6 weeks, subsequent attacks can be controlled by applying a drench; *Aphelandra* and maidenhair fern are liable to damage.

Diflubenzuron ('Dimilin') in wettable powder form can either be added during mixing or applied as a drench. The chemical is more active in young sphagnum peats, H2 on the von Post scale, than it is in the more humified sphagnum or sedge peats. To avoid phytotoxicity occurring with *Poinsettia* and *Begonia*, the rate of application in mixes made from young sphagnum peats should only be about half the rate used with the more decomposed peats. (See also § 2.2.)

The larvae of the vine weevil attacks some glasshouse and nursery stock plants, e.g. *Cyclamen* and *Rhododendron*. For ornamental plants Aldrin dust can be incorporated into the mix at $1.6\,\mathrm{kg\,m^{-3}}$ ($42\,\mathrm{oz\,yd^{-3}}$), on edible crops use either diflubenzuron ('Dimilin'), or gamma-HCH ('Lindane') as a drench.

CHAPTER FOURTEEN

Plant containers, modules and blocks

Plant containers are made from a wide range of materials, e.g. clay, plastic, metal, peat and paper; also many plants are raised in units of compressed media without a container, these are known as peat or soil blocks. It is not intended to discuss the merits of each container, only the physical and chemical interactions that occur between the container and the media which affect plant growth. The literature relating to container type has been reviewed (Bunt 1960).

14.1 CLAY v. PLASTIC POTS

The essential difference between these two types of container is one of porosity. Plastic pots are non-porous, whereas clay pots can have varying degrees of porosity, depending upon the manner in which they are made. Porous clay pots differ from plastic pots in three main ways:

(a) water is lost by evaporation from the clay wall, hence the medium can be drier;
(b) the latent heat of evaporation from the clay wall results in a reduction in the temperature of the medium;
(c) the flow of water from the medium into the clay wall results in the loss of some nutrients to the plants.

14.1.1 *Water loss*

The comparative water loss from clay and plastic pots can best be determined from fallow pots. This overcomes any effect which the pot may have upon the amount of growth made by the plant, which in turn affects the rate of water loss. Clay pots lose water by evaporation from the media surface and also from the clay walls, whereas plastic pots lose water only by evaporation from the media surface. The relative importance of water lost from these two sources was determined by Bunt & Kulwiec (1971) and was found to vary with the

Table 14.1 Mean evaporative water loss from fallow plastic and clay pots.

Container type	Water loss ($g \, day^{-1}$)
Winter	
plastic	8.1
clay	14.9
clay wall only	6.3
Summer	
plastic	22.8
clay	34.4
clay wall only	8.7

season (Table 14.1). In winter, the mean daily evaporative loss from 10.6 cm ($4\frac{1}{4}$ in) diameter plastic pots was approximately 8.1 g, and from comparable clay pots it was 14.9 g, a difference of 6.8 g or an increase of 85% by the clay pots. In summer, the mean daily losses were 22.8 g for plastic pots and 34.4 g for clay pots, a difference of 11.6 g or an increase of 50% by the clay pots. It was found that, in winter, approximately 42% of the total amount of water lost by evaporation from the clay pots occurred through the pot wall whereas in summer the loss through the clay wall was reduced to 25% of the total loss.

In practical terms this means that, although the *actual rate* of drying out of clay pots will always be greater than that of plastic pots, *the relative rate at which clay pots dry out will be greater in winter than it is in the summer*. Plants grown in plastic pots in winter will, therefore, be much more susceptible to over-watering and to waterlogged conditions than will those grown in clay pots. For this reason, it is of more importance to use a mix which drains freely and has a high air-filled porosity when growing in plastic pots. The beneficial effect of using porous containers in winter, when growing under excessively wet conditions caused either by the mix having a poor physical structure with low aeration or by watering the pots too frequently, is shown in Figure 14.1. Although it is impossible to separate entirely other factors such as media temperature and salinity, it can be concluded from other treatments in the experiment that the effect shown in Figure 14.1 is predominantly due to excess water and lack of aeration. *Primula* and *Cyclamen* are two examples of plants which are sensitive to over-watering and wet conditions in winter. It should be noted that, although the porous clay pot will allow the passage of water through the pot wall, it is not permeable to air. The improvement in media aeration when using clay pots is solely due to the loss of water by evaporation and its replacement with air drawn down from the media surface.

Richards (1974) has examined the effect of the number and size of the drainage holes on the amount of water retained in media in plastic containers. He calculated that with only a minimal water pressure of 1 mm, an unobstructed drainage hole of 6.35 mm diameter would pass 1732 cm^3 of water per hour (i.e. a $\frac{1}{4}$ in hole will pass three pints of water per hour). In practice, the drainage holes are not completely unobstructed and the rate

Figure 14.1 Reduced growth in plastic pots resulting from over-watering. When the frequency of watering was reduced and the physical structure of the mix improved, plants grown in plastic pots were not inferior to those grown in clay pots.

at which water will pass through the hole will be somewhat less than this. The most effective position of the drainage hole was found to be in the base of the pot. When the same number of drainage holes were positioned around the side of the pot so that the lower edge of each hole was directly at the base, drainage was less effective. Pots which had 'crocks' placed in the base, i.e. pieces of plastic to serve as coarse drainage material (which was formerly the practice amongst private gardeners), actually retained more water than those pots without crocks. The crocks broke the continuity of the water film between the media particles and the bench surface and so impeded the flow of water from the pot.

It was concluded that the number and size of holes in plastic pots are not normally factors which limit the drainage. The structure of the medium and the type of surface upon which the pots are stood, i.e. an open-type bench or a sand or gravel surface, are the most important factors in allowing the water to drain from the pots (see also § 3.4 and 10.2).

14.1.2 Temperature

When water evaporates, heat is lost, the latent heat of water evaporation being 2443 J g^{-1} at 20 °C. The increased rate of evaporation from clay pots can therefore be expected to result in a lower media temperature than occurs in plastic pots. Media temperatures in clay and plastic pots were measured under a wide range of environmental conditions by Bunt & Kulwiec (1970). The difference in temperature between the two types of container varied with the rate of water evaporation and ranged from about 1 °C (1.8 °F) by night in winter to 4 °C (7.2 °F) by day in summer.

Although the temperature difference of 1 °C in winter is small and might therefore be expected to have little effect on plant growth, experiments have shown that, in winter, plants in plastic pots made about 15% more growth than those in clay pots. This was because the temperature of the media was often below the optimal level for plant growth and it then

Figure 14.2 In winter, plants in plastic pots make more growth than those in clay pots because of the higher soil temperature, provided that there are no adverse effects of salinity and over-watering. In summer, the temperature in plastic pots may be too high and so restrict plant growth.

became a limiting factor. It is not generally realized that under radiation conditions on a cold, clear winter's night, the temperature of the pots falls steadily during the night, and by dawn the media temperature can be as much as 5.5 °C (10 °F) below that of the air temperature of the glasshouse, i.e. at a controlled air temperature of 15.6 °C (60 °F) the media temperature can be only 10 °C (50 °F). A typical example of the type of container affecting plant growth by way of media temperature is shown in Figure 14.2. In this experiment, both plants had been watered by tensiometer readings and salinity levels were low; the difference in growth was due to the effect of the container on the media temperature.

In summer, however, the temperature of the media can be appreciably above the optimal value; temperatures of up to 42 °C (108 °F) have been recorded in experiments at the Glasshouse Crops Research Institute in mid-summer. At such high temperatures, growth can be adversely affected and evaporative cooling becomes beneficial; plants in clay pots then make more growth than those in plastic pots. The temperature of the media is also influenced by the colour of the pot; clay pots are terra-cotta in colour whereas plastic pots are available in a range of colours. Measurements made with thermocouples have shown that black and terra-cotta plastic pots have virtually the same media temperature, whereas the temperature of white plastic pots can be as much as 4 °C (7.2 °F) lower than that of the other colours. Often the daytime temperature of the media in white plastic pots will be approximately the same as that in clay pots. This is because the lower rate of solar heat absorption by the white plastic pot is approximately equal to the evaporative cooling effect of the clay pot. By night, however, the colour of the pot has no effect on its temperature and the clay pot is then cooler than any of the plastic pots. Claims that pots made of expanded polystyrene have a higher temperature by night than

274

pots made of high density polystyrene have not been confirmed by experimental measurements.

Pot colour can also affect the plant by the amount of light that is transmitted through the plastic wall. Pots made of white or light green plastic can transmit sufficient light to adversely affect root growth and encourage the growth of algae, in some instances shoot growth is also affected. Plants reported as being sensitive include *Poinsettia*, *Begonia*, *Aglaeonema* and *Chamaedorea*.

14.1.3 Nutrient absorption

The third important effect of the container concerns plant nutrition. Water that moves from the media into the pot walls and is then lost by evaporation contains plant nutrients; these nutrients are deposited in the clay wall and are effectively lost to the plant. Often the build-up of nutrients in the pot wall is seen as a powdery white deposit on the outer surface of the wall. Measurement of the salinity of the media in clay pots has shown there is a high nutrient concentration in the centre of the pot and a relatively low nutrient concentration in the media adjacent to the inside wall. This loss of nutrients into the pot wall can be beneficial in the case of young seedlings grown in media having high salinity values. Under these conditions, the clay pot has an advantage over the plastic pot. For example, if tomato seedlings have been pricked out in winter into a John Innes Potting Compost No. 3 instead of into a JIP-1 compost, then the seedlings grown in clay pots will become established and grow more quickly than those in plastic pots. Conversely, when the nutrient levels in the media are not too high and the growth rate is rapid, any loss of nutrients into the pot wall is detrimental and plants grown in clay pots will show starvation symptoms and a reduced growth rate before comparable plants grown in plastic pots.

An example of the retarding effect which a high salinity has on the growth of plants in plastic pots is shown in Figure 14.3.

Figure 14.3 Using a medium with a high salinity can cause reduced growth in plastic pots. Clay pots are able to absorb some of the salts and so give better results.

14.1.4   Pot cleaning

Both clay and plastic pots should be thoroughly cleaned before being re-used, and plastic pots, because of their smooth, non-absorbent surface, are more easily washed than clay pots. In addition to the removal of media particles, all trace of fungi and bacteria which might infect subsequent crops should also be removed. Clay pots can be steam sterilized but most types of plastic pot will melt if heated to this temperature; polypropylene is one type of plastic that can be steamed.

Of the chemicals that have been tried as disinfectants, a formaldehyde soak for 10 min at a strength of 20 $cm^3 l^{-1}$ (1 pint per 6 gal) of water has given the best results (Nichols & Jordon 1972). *Because of the irritating vapour, treatment should be given in the open, and rubber gloves worn.* After treatment, the pots should be stacked and covered with a plastic sheet for 24 hrs, then uncovered and hosed down with clear water at intervals for a few days until the smell of formaldehyde has disappeared. If the pots are allowed to dry out too quickly, paraformaldehyde is formed and this can be phytotoxic.

14.2   PAPER AND PEAT POTS

Pots made of paper and compressed peat have the advantage that the plant need not be removed from the pot before planting and no labour is required in gathering up the empty pots and cleaning them before re-use.

Paper pots can be subdivided into two groups, those which have been treated to make them waterproof and those which have not. Waterproofing prevents the paper from absorbing water, thereby allowing it to retain its strength and to resist bacterial decomposition. Having a dry wall with no evaporative cooling effect, the media temperature in this type of pot is similar to that in plastic pots. Untreated paper pots absorb water and media temperature in them responds in a similar way to that in clay pots. Pots made of compressed peat, which are usually made with a binding agent to increase their tensile strength, absorb water and behave in a similar way to clay pots with respect to temperature and nutrient absorption. Plant roots are, however, able to penetrate the compressed peat wall.

One important factor regarding the use of pots made of paper, peat and similar materials, is their susceptibility to bacterial decomposition and the subsequent reduction in the amount of mineral nitrogen available to the plant. Some materials decompose much more readily than others and the temporary lock-up of mineral nitrogen can result in a reduced growth rate by comparison with plants grown in plastic or clay pots. Provided that this situation can be recognized at an early stage, it can be corrected with a nitrogen feed. Most pots made of compressed peat now have some mineral nitrogen added during their manufacture to allow for this phenomenon.

When plants in paper and peat pots are planted into the field, it is important to see that the surrounding soil is kept moist. If the paper is allowed to dry out, the roots may have some difficulty in penetrating the pot wall and growing into the soil, resulting in a severe check to plant growth. Watering with a surfactant solution (§ 3.6.2) before transplanting will overcome this problem.

14.3   MODULES AND BLOCKS

In the past decade there has been a dramatic increase in the number of vegetable transplants raised in modules or blocks rather than by the traditional seed-bed or bare-rooted method. An estimated 1000 million transplants were produced in the USA in 1981 and 480 million in the UK in 1984. The principle is not new, blocks of peat- or mineral-based potting mixes were used in the 1920s in the USA and in Europe in the 1940s–50s. Growers also use the system for raising lettuce plants and rooting AYR chrysanthemums in greenhouses; greater uniformity and earlier maturity are obtained. A new variation of the principle is the very small modules, known as 'plugs', used for bedding plant production. Examples of plants raised by the different systems are given in Figure 14.4.

The terms 'module' and 'block' are sometimes used synonymously. In this chapter 'module' is used where the mix is loosely placed in a tray having numerous honeycomb cells. These may be shaped as wedges or inverted pyramids in a polystyrene block and have relatively large drainage holes, e.g. 'Speedling', or as round cells in a thin rigid plastic tray, e.g. 'Hassy'. The term 'plug' is reserved for very small modules, and 'block' is used for units of compressed media, either peat or soil mixes, which are not supported by a container.

Modules

The volume of mix used to raise each plant is small by normal container sizes, ranging from 6 cm^3 for plugs to 80 cm^3; for cauliflower transplants 24 cm^3 is the preferred size.

Physical aspects   Mixes can include peat, vermiculite, polystyrene beads, calcined clay and pine bark. An essential requirement is they should be fairly fine and 'flow' easily to fill the cell uniformly. As the cells are shallow, about 2–5 cm deep, it might be reasoned that a very coarse mix was required to avoid overwet conditions (Ch. 3). Factors that help in growing in such shallow units are:

(a) the cells are completely filled with media, there is no rim to hold free water over the media;
(b) the volume of mix for each plant is small, once established the plants soon create some air-filled porosity by transpirational loss of water.

As the media volume is so small, transplant size is often directly related to the cell volume. Large, soft growing plants are not acceptable as under

Figure 14.4 Examples of plants in modules and blocks; (a) cauliflowers raised in modules, 'Speedling' (left) and 'Hassy' (right), (b) geranium cutting in 'Jiffy 7', (c) celery in a peat block.

(b)

(c)

hot, dry conditions they do not always survive transplanting as well as harder plants.

The shape of the cell, i.e. inverted pyramid or cylinder, is largely decided by economics, choice of mechanical transplanting system or user preference rather than by specific requirements of plants. Roots in wedge-shaped cells are 'air pruned', which encourages secondary root development; plants in cylindrical cells form root balls similar to those of plants in large containers. The latter type of root system is sometimes preferred for onions, whereas for sweetcorn the wedge-type module is preferred to block-raised plants.

Chemical aspects The diverse range of crops, growing environments and systems of management makes the use of various fertilizer concentrations and liquid feeds necessary. Generally, low levels of nitrogen and potassium in the mix are preferred; high rates of nitrogen can cause malformation, e.g. bent stems in Brussels Sprouts. In the UK typical nutrient concentrations (mg l^{-1}) in module mixes are:

N	P	K
100	120	150

Growth is controlled by liquid feeding, which is varied to obtain different types of plants (Hiron & Symonds 1984). A feed of 50 mg l^{-1} nitrogen, 100 mg l^{-1} potassium once per week will 'hold' plants until required for transplanting. Sturdy plants with low growth rates can be obtained with either this feed at every watering or 100 mg l^{-1} nitrogen, 200 mg l^{-1} potassium at 2–3 applications per week. If the plants have been held on a low feed, active growth can be stimulated before transplanting by feeding with 200 mg l^{-1} nitrogen, 200 mg l^{-1} potassium.

A mix suitable for rooting chrysanthemum cuttings in modules prior to planting into their flowering positions has similar phosphorus and potassium concentrations to the GCRI Potting Mix I, but the nitrogen is reduced by omitting the ammonium nitrate to give:

	N	P	K
mg l^{-1}	100	120	290

Liquid feeds for module plants are varied with the media and the climate. Smith (1984) found a feed of 200 mg l^{-1} nitrogen, 25 mg l^{-1} phosphorus, 150 mg l^{-1} potassium was necessary when growing in a mix of fine milled (15 mm screen) South African pine bark.

Plugs

The term 'plug' is used here for cells of less than 10 cm^3, in the USA it is sometimes used for a range of sizes, e.g. 648 cells per flat for bedding plants to 72–96 for geranium and tomato production. Small cells are loosely filled by brushing in a fine, moistened mix. Coarse media will not give uniform filling and very fine media cause poor aeration in shallow layers; a typical desorption curve for a plug mix based on a fine grade peat

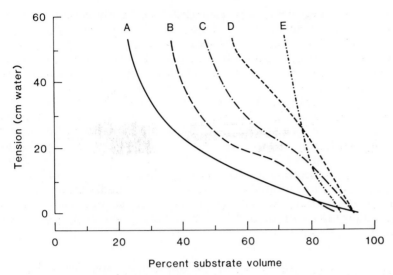

Figure 14.5 Desorption or water release curves of a plug mix and various units of compressed peat, compared with a young sphagnum peat. Note the absence of large pores that drain at low tensions in the plug mix B in comparison with the fibrous peat A.
Key: A young fibrous peat, uncompressed;
    B flux mix, uncompressed;
    C Jiffy No. 9, compressed;
    D Jiffy No. 7, compressed;
    E peat block, made from compressed humified sphagnum.

is included in Figure 14.5. Some growers prefer to cover the seeds with a coarser grade of vermiculite to give better seed germination and establishment.

Plugs are usually direct sown and low nutrient levels are required for even and quick germination of some flower crops, e.g. *Salvia*. A suitable fertilizer rate is: single superphosphate 0.75 kg m^{-3} (1 lb 5 oz yd^{-3}) and potassium nitrate 0.375 kg m^{-3} (10½ oz yd^{-3}). This gives:

	N	P	K
mg l^{-1}	52	60	142

Tomato plants and brassicas germinate well at higher nutrient levels.

Blocks

Physical aspects Blocks are made by mechanically compressing media into cubes. Compression increases the bulk density, reduces aeration and the volume of water available at low tensions. Blocks made from young fibrous sphagnum peat usually lack tensile strength and are easily damaged during irrigation and transplanting. The more humified black sphagnum and sedge peats have a finer structure and bind well to give stable blocks, early root development, however, may be slower in these media. Other forms of compressed peat are 'Jiffy 7' and 'Jiffy 9'. The former is a young sphagnum peat (H2–H3), highly compressed and encased in a weak net of polyethylene and polypropylene. When placed in a shallow tray of water, the compressed unit, having an original volume of 16.3 cm^3, expands five fold to 83.2 cm^3. The average bulk density is 0.102 g cm^{-3}, this is considerably lower than the BD of peat blocks, which varies with the operator and the peat type, but which will be about 0.171 g cm^{-3}. The smaller Jiffy 9 has an expanded volume of about 36.3 cm^3 and a BD of 0.107 g cm^{-3}; it has no supporting net, the peat being held by a bituminous binding agent. Typical desorption curves for Jiffy 7 and 9 and peat blocks are included in Figure 14.5. Aeration in the mechanically compressed peat block is poor by potting media standards, but the absence of a retaining wall with a rim to hold free water over the media means that less water passes into the block; also the large surface area means greater evaporation. The Jiffy units have lower bulk densities and are better aerated.

Chemical formulation The rates of applied phosphorus and potassium are often similar to those of the GCRI Potting Mix I (§ 8.8), i.e. 120 mg l^{-1} phosphorus and 290 mg l^{-1} potassium. The total nitrogen added ranges from 255 mg l^{-1} (103 mg l^{-1} as nitrate nitrogen and 152 mg l^{-1} as urea-formaldehyde) in winter, to 366 mg l^{-1} (138 mg l^{-1} nitrate nitrogen and 228 mg l^{-1} urea-formaldehyde) for high nitrogen type blocks in summer. Urea-formaldehyde is converted to nitrates more rapidly in black sphagnum and sedge peats than it is in young sphagnum peats. Consequently salinity levels in the blocks can rise sufficiently rapidly to impair lettuce germination and seedling growth; nitrogen rates for lettuce should not exceed 100–150 mg l^{-1}, and preferably be in inorganic form. It must also be remembered that as the peat blocks are made from compressed fertilized medium, the *effective* salinity will be correspondingly greater than if the medium was at a lower BD in a container.

Insecticides such as 'Birlane' (chlorfenvinphos) and 'Dursban' (chlorpyrifos) can be incorporated into the block for control of cabbage root fly (Saynor & Ann 1983). Fungicides such as 'Aaterra' (Etridiazole), 'Filex' (Propamocarb) and 'Basilex' (Tolclofos-methyl) can be added for controlling *Pythium*, *Phytophthora* and *Rhizoctonia*. No other chemical should be added to a mix containing calomel.

Planting into field soils

The speed with which transplants are able to establish in the field is largely determined by the water stress (Costello & Paul 1975, Kratky *et al.* 1980, Nelms & Spomer 1983). Transplants lose water by drainage into the field soil as well as by transpiration; consequently the water requirement of plants after transplanting, and before they become established, is greater than that of similar size plants remaining in containers. Irrigating the primary rooting medium, i.e. module, block or ex-container root ball, is more effective in obtaining good establishment than irrigating the surrounding field soil.

Tomato module (Grow Bags)

In 1984, approximately 50% of greenhouse tomatoes in the UK were grown in peat modules. The module normally consists of 20 l of fertilized peat, either sedge or humified sphagnum, placed in a polythene bag. Two plants are grown in each module and the yield can be up to 15 kg (30 lb) per plant. Only a small amount of the total nutrients required can therefore be supplied in the base fertilizer. If modules were made by the grower, typical base fertilizer rates would be:

$kg\,m^{-3}$		yd^{-3}
1.75	superphosphate	3 lb
0.87	potassium nitrate	1 lb 8 oz
0.44	potassium sulphate	12 oz
4.2	ground limestone	7 lb
3.0	Dolomite limestone	5 lb
0.4	Frit 253A	10 oz

This gives an added nutrient content of:

	N	P	K
$mg\,l^{-1}$	120	148	510

Additional slow-release nitrogen as 0.44 kg m^{-3} urea-formaldehyde (167 mg l^{-1} of nitrogen) is sometimes included. If a slow-release phosphorus fertilizer is required, magnesium-ammonium-phosphate ('MagAmp' or 'Enmag') at 1.5 kg m^{-3} is added. Frit 253A with its higher boron content is preferred to Frit WM255 for tomato plants. One variation of the tomato module, which is made from a loose, moist peat, is the 'Vapo' peat growing board. This is a sheet of fertilized and compressed, dry, young sphagnum peat (about 28% moisture on the moist weight basis), loosely wrapped in a polythene bag. When required for use the compressed peat is expanded by adding water; the principal being similar to the 'Jiffy 7'.

Appendices

Table A.1 Metric conversions.

	To convert	To	Multiply by
mass	pounds	kilograms	0.4536
	ounces	grams	28.3495
	kilograms	pounds	2.2046
	grams	ounces	0.0353
length	yards	metres	0.9144
	inches	centimetres	2.54
	metres	yards	1.0936
	centimetres	inches	0.3937
area	square yards	square metres	0.8361
	square inches	square centimetres	6.4516
	square metres	square yards	1.1959
	square centimetres	square inches	0.1550
pressure	pounds in^{-2}	atmospheres	0.0680
	pounds in^{-2}	kg m^{-2}	703.0696
	atmospheres	pounds in^{-2}	14.6959
	kg m^{-2}	pounds in^{-2}	0.0014
	atmospheres	KPa	101.325
volume and capacity	cubic yards	cubic metres	0.7645
	cubic feet	litres	28.3168
	bushels	cubic metres	0.0364
	gallons (Imperial)	litres	4.5461
	gallons (US)	litres	3.7854
	fluid ounces (Imperial)	millilitres	28.4122
	fluid ounces (US)	millilitres	29.5727
	cubic metres	cubic yards	1.3080
	litres	cubic feet	0.0353
	cubic metres	bushels	27.4967
	litres	gallons (Imperial)	0.2199
	litres	gallons (US)	0.2642
	millilitres	fluid ounces (Imperial)	0.0352
	millilitres	fluid ounces (US)	0.0338
density and concentration	lb yd^{-3}	kg m^{-3}	0.5932
	oz yd^{-3}	g m^{-3}	37.0797
	oz bushel^{-1}	g l^{-1}	0.7795
	oz gal^{-1} (Imperial)	g l^{-1}	6.2361
	oz gal^{-1} (US)	g l^{-1}	7.4891
	kg m^{-3}	lb yd^{-3}	1.6855
	g m^{-3}	oz yd^{-3}	0.0269

Continued on page 285

	To convert	To	Multiply by
	$g\,l^{-1}$	$oz\,bushel^{-1}$	1.2829
	$g\,l^{-1}$	$oz\,gal^{-1}$ (Imperial)	0.1603
	$g\,l^{-1}$	$oz\,gal^{-1}$ (US)	0.1335
energy	British thermal unit	kilojoules	1.0551
	$BTU\,lb^{-1}$	$joules\,kg^{-1}$	2326
	$BTU\,ft^{-3}$	$kilojoules\,m^{-3}$	37.2589
	$BTU\,ft^{-3}\,°F$	$kilojoules\,m^{-3}\,°C$	67.0661
	kilowatt hours	megajoules	3.6

Table A.2 Imperial and US capacity measures.

	Measure		Equivalent to
1	Imperial gal	1.2009	US gal
1	Imperial gal	4.5459	litre
1	Imperial gal	160	fl. oz
1	Imperial fl. oz	0.9607	US fl. oz
1	US gal	0.8326	Imperial gal
1	US gal	3.7853	litre
1	US gal	128	fl. oz
1	US fl. oz	1.0408	Imperial fl. oz
1	Imperial bushel	1.0321	US bushels
1	US bushel	0.9689	Imperial bushel
1	Imperial bushel	36.37	litre
1	US bushel	35.24	litre
1	yd^3	21.016	Imperial bushels
1	yd^3	21.71	US bushels

Table A.3 Illumination and solar radiation units.

	Unit		Equivalent to
1	$lumen\,ft^{-2}$	1	foot candle
1	$lumen\,m^{-2}$	1	lux
1	foot candle	10.76	lux
1	$cal\,cm^{-2}\,min^{-1}$	697.6	$joules\,m^{-2}\,sec^{-1}$
		697.6	$watts\,m^{-2}$
7500	$lumens\,ft^{-2}*$	c. 1	$cal\,cm^{-2}\,min^{-1}$

* The relationship of lumens to calories changes slightly with the time of the year and the amount of cloud. The figure of 7500 lumens per calorie is an average value.

Table A.4 Atomic weights.

hydrogen	H	1.00	sulphur	S	32.066
boron	B	10.82	chlorine	Cl	35.457
carbon	C	12.01	potassium	K	39.096
nitrogen	N	14.008	calcium	Ca	40.08
oxygen	O	16.00	manganese	Mn	54.93
sodium	Na	22.997	iron	Fe	55.85
magnesium	Mg	24.32	copper	Cu	63.54
aluminium	Al	26.97	zinc	Zn	65.38
phosphorus	P	30.98	molybdenum	Mo	95.95

Table A.5 Formulae and molecular weights of some commonly used chemicals.

aluminium sulphate	$Al_2(SO_4)_3 \cdot 18H_2O$	666
ammonium nitrate	NH_4NO_3	80
diammonium phosphate	$(NH_4)_2HPO_4$	132
monoammonium phosphate	$NH_4H_2PO_4$	115
ammonium sulphate	$(NH_4)_2SO_4$	132
calcium carbonate	$CaCO_3$	100
calcium hydroxide	$Ca(OH)_2$	74
calcium oxide	CaO	56
calcium nitrate	$Ca(NO_3)_2 \cdot 4H_2O$	236
calcium nitrite	$5Ca(NO_3)_2 \cdot NH_4NO_3 \cdot 10H_2O$	1080
magnesium sulphate	$MgSO_4 \cdot 7H_2O$	246
potassium chloride	KCl	74
potassium dihydrogen phosphate	KH_2PO_4	136
potassium nitrate	KNO_3	101
potassium sulphate	K_2SO_4	174
urea	$CO(NH_2)_2$	60

Table A.6 Formulae and molecular weights of some chemicals commonly used to supply microelements.

Chemical	Formula	Molecular weight	Amount of microelement
borax	$Na_2B_4O_7 \cdot 10H_2O$	382	11.3% B
boric acid	H_3BO_3	62	17.5% B
Solubor	$Na_2B_8O_{13} \cdot 4H_2O$	412	20.5% B
copper sulphate	$CuSO_4 \cdot 5H_2O$	249	25.5% Cu
ferrous sulphate*	$FeSO_4 \cdot 7H_2O$	278	20.1% Fe
manganese sulphate	$MnSO_4 \cdot 4H_2O$	223	24.6% Mn
sodium molybdate	$Na_2MoO_4 \cdot 2H_2O$	242	39.7% Mo
zinc sulphate	$ZnSO_4 \cdot 7H_2O$	287	22.7% Zn

* Iron chelates range from c. 14% iron for EDTA-Fe to c. 6% for EDDHA-Fe.

Table A.7 Chemical gravimetric conversions.

To convert	To	Multiply by
NH_4	N	0.778
NO_3	N	0.226
N	NH_4	1.285
N	NO_3	4.427
P_2O_5	P	0.436
PO_4	P	0.326
P	P_2O_5	2.291
P	PO_4	3.066
K_2O	K	0.830
K	K_2O	1.205
$CaCO_3$	Ca	0.400
CaO	Ca	0.714
Ca	$CaCO_3$	2.497
Ca	CaO	1.399
$MgCO_3$	Mg	0.288
MgO	Mg	0.603
Mg	$MgCO_3$	3.467
Mg	MgO	1.657

Table A.8 Temperature conversions.

To convert °F to °C: $(°F - 32) \times \frac{5}{9}$
To convert °C to °F: $(°C \times \frac{9}{5}) + 32$

°F	°C	°F	°C
32	0	75	23.9
40	4.4	80	26.7
45	7.2	100	37.8
50	10.0	120	48.9
55	12.8	140	60.0
60	15.6	160	71.1
65	18.3	180	82.2
70	21.2	200	93.3

An increase of 1 °F = 0.555 °C
An increase of 1 °C = 1.8 °F

Bibliography

Aaron., J. R. 1982. *Conifer bark: its properties and uses.* Forestry Commission Forest Record 110. London: HMSO.

Adams, P. & D. M. Massey 1984. Nutrient uptake by tomatoes from recirculating solutions. *Proc. 6th Int. Congress on Soilless Culture,* Lunteren, 71–9.

Aendekerk, T. G. L. *Gebreksziekten in boomkwekerijgewassen.* Proefstation voor de boomkwekerij C. A. D. voor de boomkwekerij, Boskoop.

Aguila, J. F. & F. X. Martinez 1980. Outdoor hydroponics in hanging growing boxes located around big buildings. *Proc. 5th Int. Congress on Soilless Culture,* 347–64. Wageningen: ISOSC.

Airhart, D. L., N. J. Natarella & F. A. Pokorny 1978a. The structure of processed pine bark. *J. Am. Soc. Hort. Sci.* **103**, 404–8.

Airhart, D. L., N. J. Natarella & F. A. Pokorny 1978b. Influence of initial moisture content on the wettability of a milled pine bark medium. *HortScience* **13**, 432–4.

Aldrich, R. A. 1976. Heat treatment of growing media for floriculture and nursery production. *Penn. Flo. Gro. Bull.* **291**, 1–10.

Aldrich, R. A., J. W. White & P. E. Nelson 1972. Aerated steam. II Engineering requirements, design and operation of systems for aerated steam treatment of soil and soil mixtures. *Penn. Flo. Gro. Bull.* **253**, 3–7.

Allen, R. C. 1943. Influence of aluminium on the flower colour of *Hydrangea macrophylla* D.C. *Contributions from Boyce Thompson Inst.* **13**, 221–42.

Allison, F. E. & R. M. Murphy 1963. Comparative rates of decomposition in soil of wood and bark particles of several species of pines. *Soil Sci. Soc. Am. Proc.* **27**, 309–12.

Allison, F. E., R. M. Murphy & C. J. Klein 1963. Nitrogen requirements for the decomposition of various kinds of finely ground woods in soils. *Soil Sci.* **96**, 187–91.

Alvey, N. G. 1955. Adapting John Innes composts to grow *Ericas. J. R. Hort. Soc.* **80**, 376–81.

Alvey, N. G. 1961. Soil for John Innes composts. *J. Hort. Sci.* **36**, 228–40.

Arnold Bik, R. 1970. Nitrogen, salinity, substrates and growth of *Gloxinia* and *Chrysanthemum. Mededeling* no. 3, Wageningen: Centre for Agricultural Publishing and Documentation.

Arnold Bik, R. 1972. *Influence of nitrogen, phosphorus and potassium rates on the mineral composition of the leaves of the Azalea variety Ambrozius.* Colloquium Proc. no. **2**, 99–102. Henley-on-Thames: Potassium Institute.

Arnold Bik, R. 1973. Some thoughts on the physical properties of substrates with special reference to aeration. *Acta Horticulturae* **31**, 149–60.

Arnold Bik, R. 1983. Substrates in floriculture. *Proc. XXI Int. Hort. Congress,* Hamburg *29/8 to 4/9/82.* **2**, 811–22.

Arnon, D. I. & C. M. Johnson 1942. Influence of hydrogen-ion concentration on the growth of higher plants under controlled conditions. *Plant Physiol.* **17**, 525–39.

Asen, S. & C. E. Wildon 1953. Nutritional requirements of greenhouse chrysanthemums growing in peat and sand. *Q. Bull. Mich. Agric. Expl Stn* **36**, 24–9.

Baker, K. F. (ed.) 1957. The U.C. system for producing healthy container-grown plants. *Calif. Agric. Expl Stn Manual* **23**.

Baker, K. F. 1967. Some microbiological effects of soil treatment with steam and chemicals. *Proc. Washington State University's Greenhouse Growers Inst.* 20–2 June.

Bar-Akiva, A. 1964. Visible symptoms and chemical analysis vs biochemical indicators as a means of diagnosing iron and manganese deficiencies in citrus plants. In *Plant analysis and fertilizer problems*, C. Bould, P. Prevot & J. R. Magness (eds), vol. 4, 9–25. Geneva, New York: Am. Soc. Hort. Sci.

Batson, F. 1972. *Azalea*. In the *Ball Red Book*, 12th edn, 195–214. Chicago: G.T. Ball.

Beardsell, D. V. & D. G. Nichols 1982. Wetting properties of dried-out nursery container media. *Scientia Hort.* **17**, 49–59.

Beardsell, D. V., D. G. Nichols & D. L. Jones 1979a. Physical properties of nursery potting-mixtures. *Scientia Hort.* **11**, 1–8.

Beardsell, D. V., D. G. Nichols & D. L. Jones 1979b. Water relations of nursery potting-media. *Scientia Hort.* **11**, 9–17.

Bennett, A. C. & F. Adams 1970. Concentration of NH_3 (aq.) required for incipient NH_3 toxicity to seedlings. *Proc. Am. Soc. Soil Sci.* **34**, 259–63.

Bernstein, L. 1963. Osmotic adjustment of plants to saline media. II Dynamic phase. *Am. J. Bot.* **50**, 360–70.

Besford, R. T. & J. L. W. Deen 1977. Peroxidase activity as an indicator of the iron status of conifers. *Scientia Hort.* **7**, 161–9.

Biamonte, R. L. 1977. Water soluble fertilization. *Florida Foliage Grower* **14**(4), 1–3.

Bingham, F. T. 1959. Micronutrient content of phosphorus fertilisers. *Soil Sci.* **88**, 7–10.

Bingham, F. T. & M. J. Garber 1960. Solubility and availability of micronutrients in relation to phosphorus fertilisation. *Soil Sci. Soc. Am. Proc.* **24**, 209–13.

Birch, P. D. W. & D. J. Eagle 1969. Toxicity to seedlings of nitrite in sterilised composts. *J. Hort. Sci.* **44**, 321–30.

Bjerre, H. 1983. Pot plant growing on a capillary mat covered with perforated polyethylene foil. *Acta Horticulturae* **133**, 161–4.

Blomme, R. & T. G. Piens 1969. *Kunstmatige bodems en bemesting van Azalea.* [Artificial soils and manuring of Azaleas]. B.V.O. mededelingen. Bedrijfsvoorlichtinchtingsdienst voor de Tuinbouw in de Provincie oost. Vlaanderen VZWD no. 51.

Boertje, G. A. 1980. Results of liquid feeding in the production of bedding plants. *Acta Horticulturae* **99**, 17–23.

Boertje, G. A. 1984. Physical laboratory analyses of potting composts. *Acta Horticulturae* **150**, 47–50.

Boggie, R. & R. A. Robertson 1972. Evaluation of horticultural peat in Britain. *Proc. 4th Int. Peat Congress* **3**, 185–92.

Bollen, W. B. 1953. Mulches and soil conditions, carbon and nitrogen in farm and forest products. *J. Agric. Food Chem.* **1**, 379–81.

Boodley, J. W. 1981. Charting one's way through the maze of soil mixes. *Florists' Review*, **169**(4383), 20–1.

Boodley, J. W. & R. Sheldrake, Jr 1972. *Cornell Peat-lite mixes for commercial plant growing*. Information Bull. no. 43. NY College of Agric., Cornell Univ., New York.

Boxma, R. 1981. Effect of pH on the behaviour of various iron chelates in sphagnum (moss) peat. *Comm. in Soil Sci. and Plant Analysis* **12**, 755–63.

Bragg, N. C. & B. J. Chambers 1988. The interpretation and advisory application of compost air-filled porosity (AFP) measurements. *Acta Horticulturae* (in press).

Branson, R. L., R. H. Sciaroni & J. M. Rible 1968. Magnesium deficiency in cut-flower chrysanthemums. *Calif. Agric.* **22**(8), 13–14.

Broschat, K. & H. M. Donselman 1985. Extractable Mg, Fe, Mn, Zn and Cu from a peat-based container medium amended with various micronutrient fertilizers. *J. Am. Soc. Hort.* **110**, 196–200.

Broschat, T. K. & G. E. Fitzpatrick 1980. Removal of irrigation water residues from foliage of ornamental plants. *Proc. Fla State Hort. Soc.* **93**, 205–7.

Brown, E. F. & F. A. Pokorny 1975. Physical and chemical properties of media composed of milled pine-bark and sand. *J. Am. Soc. Hort. Sci.* **100**, 119–21.

Bugbee, G. J. & C. R. Frink 1983. *Quality of potting soils.* Bulletin 812, Connecticut Agricultural Experiment Station.

Bunt, A. C. 1954a. Steam pressure in soil sterilisation. I In bins, *J. Hort. Sci.* **29**, 89–97.

Bunt, A. C. 1954b. *Steam sterilisation. Steam–air mixture.* Ann. Rep. John Innes Hort. Inst., 28.

Bunt, A. C. 1956. An examination of the factors contributing to the pH of the John Innes Composts. *J. Hort. Sci.* **31**, 258–71.

Bunt, A. C. 1960. *A review of the literature on plant containers and moulded blocks with special reference to the porosity of pots.* Rep. Glasshouse Crops Res. Inst., 1959, 116–25.

Bunt, A. C. 1961. Some physical properties of pot plant composts and their effect on plant growth. III Compaction. *Plant and Soil* **XV**, 228–42.

Bunt, A. C. 1963. The John Innes Composts: Some effects of increasing the base fertilizer concentration on the growth and composition of the tomato. *Plant and Soil* **19**, 153–65.

Bunt, A. C. 1971. The use of peat-sand substrates for pot chrysanthemum culture. *Acta Horticulturae* **18**, 66–74.

Bunt, A. C. 1972. The use of fritted trace elements in peat-sand substrates. *Acta Horticulturae* **26**, 129–40.

Bunt, A. C. 1973a. *Loamless substrates for pot plants. Microelement problems.* Rep. Glasshouse Crops Res. Inst., *1972*, 66–7.

Bunt, A. C. 1973b. Factors contributing to the delay in the flowering of pot chrysanthemums grown in peat-sand substrates. *Acta Horticulturae* **31**, 163–72.

Bunt, A. C. 1974a. Physical and chemical characteristics of loamless pot-plant substrates and their relation to plant growth. *Acta Horticulturae* **37**, 1954–65.

Bunt, A. C. 1974b. *Loamless substrates for potplants. Microelement supply.* Rep. Glasshouse Crops Res. Inst., 1973, 78.

Bunt, A. C. 1976a. The use of phosphorus in liquid fertilisers. *Acta Horticulturae* **64**, 93–101.

Bunt, A. C. 1976b. *Modern potting composts*, 176. London: Allen & Unwin.

Bunt, A. C. 1980. Phosphorus sources for loamless substrates. *Acta Horticulturae* **99**, 25–32.

Bunt, A. C. 1984. Physical properties of mixtures of peats and minerals of different particle size and bulk density for potting substrates. *Acta Horticulturae* **150**, 143–53.

Bunt, A. C. & P. Adams 1966a. Some critical comparisons of peat-sand and loam-based composts with special reference to the interpretation of physical and chemical analysis. *Plant and Soil* **24**, 213–21.

Bunt, A. C. & P. Adams 1966b. *Loamless composts.* Rep. Glasshouse Crops Res. Inst., 1965, 119–20.

Bunt, A. C. & Z. J. Kulwiec 1970. The effect of container porosity on root environment and plant growth. I Temperature. *Plant and Soil* **32**, 65–80.

Bunt, A. C. & Z. J. Kulwiec 1971. The effect of container porosity on root environment and plant growth. II Water relations. *Plant and Soil* **35**, 1–16.

Bunt, A. C., J. L. Paul & A. M. Kofranek 1988. The relationship of oxygen diffusion rate to the air-filled porosity of potting substrates (in press).

Bylov, V. N., N. V. Vasilyevskaya & L. P. Vavilova 1971. Physiochemical properties of vermiculite and its use in floriculture. *Byull. Gl. Bot. Sada.* **80**, 59–63.

Carncross, C. A. 1984. Wet harvesting of peat. *Proc. Int. Symp. on Peat Utilization*, Bemidji State Univ., 10–13 Oct., 1983.

Chaney, R. L., J. B. Munns & H. M. Cathey 1980. Effectiveness of digested sewage sludge compost in supplying nutrients for soilless potting media. *J. Am. Soc. Hort. Sci.* **105**, 485–92.

Cobb, G. S. & G. J. Keever 1984. Effects of supplemented N on plant growth in fresh and aged bark. *HortScience* **19**, 127–9.

Colgrave, M. S. & A. N. Roberts 1956. Growth of the azalea as influenced by ammonium and nitrate nitrogen. *Proc. Am. Soc. Hort. Sci.* **68**, 522–36.

Conover, C. A. & R. T. Poole 1982a. Influence of nitrogen source on growth and tissue nutrient content of three foliage plants. *Proc. Fla. State Hort. Soc.* **95**, 151–3.

Conover, C. A. & R. T. Poole 1982b. Florida researchers identify foliage plants susceptible to fluoride. *Florists' Review*, **170**(4409), 42, 106.

Costello, L. & J. L. Paul 1975. Moisture relations in transplanted container plants. *HortScience* **10**, 371–2.

Court, M. N., R. C. Stephen & J. S. Waid 1962. Nitrite toxicity arising from the use of urea as a fertiliser. *Nature* **194**, 1263–5.

Dänhardt, W. & G. Kühle 1959. Experiments on the most favourable peat-clay ratio in standard peat soils to be used for pot plants (in German). *Arch. Gartenb.* **7**, 157–74.

Dasberg, S. & J. W. Bakker 1970. Characterizing soil aeration under changing soil moisture conditions for bean growth. *Agronomy J.* **62**, 689–92.

Daughtry, W. 1983. Chlorination of irrigation water. *Combined Proc. Plant Prop. Soc.* **33**, 596–9.

Davies, J. N. 1957. *Steam sterilisation studies*. Rep. Glasshouse Crops Res. Inst., 1954/55, 70–9.

Dawson, J. R., R. A. H. Johnson, P. Adams & F. T. Last 1965. Influence of steam-air mixtures, when used for heating soil, on biological and chemical properties that affect seedling growth. *Ann. Appl. Biol.* **56**, 243–51.

De Boodt, M. & O. Verdonck 1971. Physical properties of peat and peat-moulds improved by perlite and foam plastics in relation to ornamental plant-growth. *Acta Horticulturae* **18**, 9–27.

De Boodt, M. & O. Verdonck 1972. The physical properties of the substrates in horticulture. *Acta Horticulturae* **26**, 37–44.

Dempster, C. D. 1958. Clay dust compost solves the loam problem. *Commercial Grower* **3245**, 569–71.

van Dijk, H. & P. Boekel 1965. Effect of drying and freezing on certain physical properties of peat. *Neth. J. Agric. Sci.* **13**, 248–60.

Eaton, F. M. 1941. Water uptake and root growth as influenced by inequalities in the concentration of the substrate. *Plant Physiol.* **16**, 545–64.

Eaton, F. M. 1942. Toxicity and accumulation of chloride and sulfate salts in plants. *J. Agric. Res.* **64**, 357–99.

English, J. E. & A. V. Barker 1983. Growth and mineral composition of tomato under various regimes of nitrogen nutrition. *J. Plant Nutrition* **6**, 339–47.

Farnham, R. S. 1969. Classification system for commercial peat. *Proc. 3rd Int. Peat Congress*, Quebec, 85–90. Ottawa: Dept Energy, Mines and Resources.

Fisher, R. A. 1926. The arrangement of field experiments. *J. Min. Agric.* (London) **33**, 503–13.

Fitzpatrick, G. 1980. Water budget determinations for container-grown ornamental plants. *Proc. Fla State Hort. Soc.* **93**, 166–8.

Flocker, W. J., J. A. Vomocil & F. D. Howard 1959. Some growth responses of tomatoes to soil compaction. *Soil Sci. Soc. Am. Proc.* **23**, 189–91.

Fortney, W. R. & T. K. Wolf 1981. Determining nutritional status, plant analysis. *Penn. Flo. Gro. Bull.* **331**, 1, 5–11.

Foster, W. J., R. D. Wright, M. M. Alley & T. H. Yeager 1983. Ammonium adsorption on a pine-bark growing medium. *J. Am. Soc. Hort. Sci.* **108**, 548–51.

Fruhstorfer, A. 1952. *Soil mixture for horticulture.* Complete specification, Pat. Spec. 670,907. London: Brit. Patent Office.

Furuta, T., T. Mock & R. Coleman 1977. Estimating the water needed for container-grown nursery stock. *Am. Nurseryman* **145**(8), 68.

Furuta, T., R. H. Sciaroni & J. R. Breece 1967. Sulphur coated urea fertilizer for controlled release on container-grown ornamentals. *Calif. Agric.* **21**(9), 4–5.

Gabriels, R. 1978. The effect of irrigation water quality on the growing medium. *Acta Horticulturae* **82**, 201–12.

Gabriels, R., H. Engles & J. G. van Onsem 1972. Nutritional requirements of young azaleas grown in peat and coniferous litter. *Symp. Int. Peat Soc.*, Helsinki.

Gammon, N. 1957. Root growth responses to soil pH adjustments made with carbonates of calcium, sodium or potassium. *Proc. Soil Crop Sci. Soc. Florida* **17**, 249–54.

Gartner, J. B., S. M. Still & J. E. Klett 1973. The use of hardwood bark as a growth medium. *Combined Proc. Plant Prop. Soc.* **23**, 222–31.

Gasser, J. K. R. & J. E. Peachey 1964. A note on the effects of some soil sterilants on the mineralisation and nitrification of soil nitrogen. *J. Sci. Food. Agric.* **15**, 142–6.

Gauch, H. G. & C. H. Wadleigh 1944. Effects of high salt concentrations on growth of bean plants. *Bot. Gaz.* **105**, 379–87.

Gehring, J. M. & A. J. Lewis 1980. Effect of hydrogel on wilting and moisture stress of bedding plants. *J. Am. Soc. Hort. Sci.* **105**, 511–13.

Geraldson, C. M. 1967. Evaluation of the nutrient intensity and balance system of soil testing. *Proc. Soil Crop Sci. Soc. Florida* **27**, 59–67.

Gouin, F. R. & C. B. Link 1982. Sulfur tested for lowering the pH of the media amended with sewage sludge. *Am. Nurseryman* **156**, 71–9.

Green, J. L. 1968. Perlite – advantages and limitations as a growth media. *Colo. Flo. Gro. Assoc. Bull.* **214**, 4–8.

Greenwood, D. J. 1969. Effect of oxygen distribution in the soil on plant growth. In *Root growth*, W. J. Whittington (ed.), 202–21. London: Butterworth.

Guttormsen, G. 1969. Accumulation of salts in the sub-irrigation of pot plants. *Plant and Soil* **31**, 425–38.

Hanan, J. J., C. Olympios & C. Pittas 1981. Bulk density, porosity, percolation and salinity control in shallow, freely draining, potting soils. *J. Am. Soc. Hort. Sci.* **106**, 742–6.

Handreck, K. A. 1983. Particle size and the physical properties of growing media for containers. *Comm. in Soil Sci. and Plant Analysis* **14**, 209–22.

Hansen, M. 1978. Plant specific nutrition and preparation of nutrient solutions. *Acta Horticulturae* **82**, 109–12.

Harbaugh, B. K. & G. J. Wilfret 1982. Correct temperature is the key to successful use of Osmocote. *Florists' Review* **170**(4404), 21–3.

Harrod, M. F. 1971. Metal toxicities in glasshouse crops. A discussion of problems encountered in advisory work on soils of pH 6.0 and above. In *Trace Elements in Soils and Crops*. Tech. Bull. no. 21, 176–92. London: HMSO.

Hatfield, J. D., A. V. Slack, G. L. Crow & H. B. Shaffer 1958. Corrosion of metals by liquid mixed fertilisers. *J. Agric. Food Chem.* **6**, 524–31.

Haynes, R. J. 1982. Leaching losses of nutrients and yield and nutrient uptake by container-grown begonia as affected by lime and fertiliser applications to a peat medium. *J. Sci. Food Agric.* **33**, 407–13.

Haynes, R. J. & K. M. Goh 1978. Evaluation of potting media for commercial nursery production of container-grown plants. IV Physical properties of a range of amended peat-based media. *N.Z. J. Agric. Res.* **21**, 449–56.

Haynes, R. J. and R. S. Swift 1986. The effects of pH and of form and rate of applied iron on microelement availability and nutrient uptake by highbush blueberry plants grown in peat or soil. *J. Hort. Sci.* **61**, 287–94.

Helling, C. S., G. Chesters & R. B. Corey 1964. Contribution of organic matter and clay to soil cation-exchange capacity as affected by the pH of the saturating solution. *Soil Sci. Soc. Am. Proc.* **28**, 517–20.

Hershey, D. R. & J. L. Paul 1982. Leaching-losses of nitrogen from pot chrysanthemums with controlled-release of liquid fertilization. *Scientia Hort.* **17**, 145–52.

Hershey, D. R., J. L. Paul & R. M. Carlson 1980. Evaluation of potassium-enriched clinoptilolite as a potassium source for potting media. *HortScience* **15**, 87–9.

Higaki, T. & R. T. Poole 1978. A media and fertilizer study in *Anthurium*. *J. Am. Soc. Hort. Sci.* **103**, 98–100.

Hillman, W. A. & H. B. Posner 1971. Ammonium ion and the flowering of *Lemna perpusilla*. *Plant Physiol.* **47**, 586–7.

Hiron, R. W. & W. Symonds 1984. *Vegetable propagation in cellular trays*. Ministry of Agriculture, Fisheries and Food, Leaflet 909.

Hoitink, H. A. J. 1980. Composted bark, a lightweight growth medium with fungicidal properties. *Plant Disease* **64**, 142–7.

Hoitink, H. A. J. & H. A. Poole 1980. Factors affecting quality of composts for utilization in container media. *HortScience* **15**, 171–3.

Holden, E. R., N. R. Page & J. I. Wear 1962. Properties and use of micronutrient glasses in crop production. *J. Agric. Food Chem.* **10**, 188–92.

Horst, R. K., S. O. Kawamoto, G. L. Schumann & M. F. Diertert 1983. Chlorosis in healthy and viroid-infected plants exposed to the steam additive diemethylaminoethanol. *Scientia Hort.* **19**, 1–8.

Jarrell, W. M., R. A. Shepherd & R. L. Branson 1979. Leachate and soil pH changes in potting mixes treated with $NaHCO_3$ and $KHCO_3$ solutions. *J. Am. Soc. Hort. Sci.* **104**, 831–4.

Johnson, C. R. 1973a. Effects of controlled release and liquid fertilisers on the growth and mineral composition of *Celosia plumosia*. *Comm. in Soil Sci. and Plant Analysis* **4**, 323–31.

Johnson, C. R. 1973b. Symptomatology and analyses of nutrient deficiencies produced on flowering annual plants. *Comm. in Soil Sci. and Plant Analysis* **4**, 185–96.

Johnson, D. R. & T. E. Bilderback 1981. *Physical properties of aged and fresh peanut hull media and their effect on Azalea growth*. Proc. SNA Res. Conf. 26th Ann. Rep.

Johnson, E. W. 1980. Comparison of methods of analysis for loamless composts. *Acta Horticulturae* **99**, 197–204.

Johnson, J. 1919. The influence of heated soils on seed germination and plant growth. *Soil Sci.* **7**, 1–104.

Johnson, M. S. 1984. Effect of soluble salts on water absorption by gel-forming soil conditioners. *J. Sci. Food Agric.* **35**, 1063–6.

Johnson, P. 1968. *Horticultural and agricultural uses of sawdust and soil amendments*. Paul Johnson, 1904 Cleveland Ave., National City, Calif. 92050 USA.

Joiner, J. N. and C. A. Conover 1965. Characteristics affecting desirability of various media components for production of container-grown plants. *Proc. Soil Crop Sci. Soc. Florida* **25**, 320–8.

Keisling, T. C. & J. A. Lipe 1981. Comparison of post-plant liming methods for quick response in a potting media. *Comm. in Soil Sci. and Plant Analysis* **12**(5), 453–60.

Kerr, G. 1983. Salinity V; Column experiments *Colo. Greenhouse Gro. Assoc. Bull.* **402**, 1–4.

Kersten, M. S. 1949. Thermal properties of soils. *Minn. Univ. Engr. Expl. Stn Bull.* **28**, 1–228.

Kivinen, E. 1980. Proposal for a general classification of virgin peat. *Proc. 6th Int. Peat Congress*, p. 47.

Kivinen, E. 1981. Utilization of peatlands in some countries. *Int. Peat Soc. Bull.* **12**, 21–30.

Kivinen, E. & P. Pakavinem 1980. Peatland areas and proportion of virgin peatlands in different countries. *Proc. 6th Int. Peat Congress*, 52–4.

Klett, J. E., J. B. Gartner & T. D. Hughes 1972. Utilization of hardwood bark in media for growing woody ornamental plants in containers. *J. Am. Soc. Hort. Sci.* **97**, 448–50.

Kofranek, A. M. & O. R. Lunt 1975. Mineral nutrition. In *Growing azaleas commercially* A. M. Kofranek & R. A. Larson (eds), 36–45. Publication 4058, Div. Agric. Sciences, Univ. Calif.

Kohl, H. C., A. M. Kofranek & O. R. Lunt 1955. Effect of various ions and total salt concentrations on *Saintpaulia*. *Proc. Am. Soc. Hort. Sci.* **68**, 545–50.

Koths, J. S. & R. Adzima 1978. Domestic vs African vermiculite for seedlings. *Connecticut Greenhouse Newsletter* **89**, 1–7.

Kratky, B. A., E. F. Cox & J. M. T. McKee 1980. Effects of block and soil water content on the establishment of transplanted cauliflower seedlings. *J. Hort. Sci.* **55**, 229–34.

Laurie, A. 1931. The use of washed sand as a substitute for soil in greenhouse culture. *Proc. Am. Soc. Hort. Sci.* **28**, 427–31.

Lawrence, W. J. C. & J. Newell 1939. *Seed and potting composts*. London: Allen & Unwin.

Lindley, J. 1855. *Theory and practice of horticulture*. London: Longman Brown Green and Longmans.

Lindsay, W. L. & W. A. Norvell 1978. Development of a DTPA soil test for zinc, iron, manganese and copper. *Soil Sci. Soc. Am. J.* **42**, 421–8.

Link, C. B., F. R. Gouin & J. B. Shanks 1983. Composted sewage sludge and soil amendment for greenhouse and woody ornamental crops. *Acta Horticulturae* **133**, 199–203.

Long, M. I. E. & G. W. Winsor 1960. Isolation of some urea-formaldehyde compounds and their decomposition in soil. *J. Sci. Food. Agric.* **11**, 441–5.

Lucas, R. E. & J. K. Davis 1961. Relationships between pH values of organic soils and availabilities of 12 plant nutrients. *Soil Sci.* **92**, 177–82.

Luit, B. van & R. Boxma 1981. Quality check of iron chelates applied to ornamental shrubs on sphagnum peat. *J. Hort. Sci.* **56**, 125–9.

Lumis, G. P. 1980. Aquatic plant compost as a growing medium. *Compost Science/Land Utilisation* July/Aug., 33–5.

Lunt, O. R., H. C. Kohl & A. M. Kofranek 1956. The effect of bicarbonate and other constituents of irrigation water on the growth of *azaleas*. *Proc. Am. Soc. Hort. Sci.* **68**, 537–44.

MAFF/ARC 1983. *Diagnosis of mineral disorders in plants*, vols 1 & 2. London: HMSO.

Markus, D. K., J. E. Steckel & J. R. Trout 1981. Micronutrient testing in artificial mix substrates. *Acta Horticulturae* **126**, 219–25.

Matkin, O. A. & F. H. Petersen 1971. Why and how to acidify irrigation water. *Am. Nurseryman* **133**, 14 & 73.

Maynard, D. N. & A. V. Barker 1969. Studies on the tolerance of plants to ammonium nutrition. *J. Am. Soc. Hort. Sci.* **94**, 235–9.

Messing, J. H. L. 1965. The effects of lime and superphosphate on manganese toxicity in steam-sterilised soil. *Plant and Soil* **23**, 1–10.

Mitchell, R. L. 1954. Trace elements in Scottish peats. *Int. Peat Symp. Section B3*, Dublin.

Morris, L. G. 1954. *The steam sterilising of soils*. National Inst. Agric. Engr. Report, no. 24.

Morris, W. C. & D. C. Milbocker 1972. Repressed growth and leaf chlorosis of Japanese holly grown in hardwood bark. *HortScience* **7**, 486–7.

Morrison, T. M., D. C. McDonald & J. A. Sutton 1960. Plant growth in expanded perlite. *N.Z. J. Agric. Res.* **3**, 592–7.

Nelms, L. R. & L. Art Spomer 1983. Water retention of container soils transplanted into ground beds. *HortScience* **18**, 863–6.

Nelson, L. E. & R. Selby 1974. The effect of nitrogen sources and iron levels on the growth and composition of Sitka spruce and Scots pine. *Plant and Soil* **41**, 573–88.

Nelson, P. V. 1969. Assessment and correction of the alkalinity problem associated with Palabora vermiculite. *J. Am. Soc. Hort. Sci.* **94**, 664–7.

Nelson, P. V. 1972. *Greenhouse Media. The use of Cofuna, Floramull, Pinebark and Styromull*. North Carolina Agric. Expl Stn Tech. Bull. no. 206.

Nelson, P. V. & J. W. Boodley, 1966. Classification of carnation cultivars according to foliar nutrient content. *Proc. Am. Soc. Hort. Sci.* **89**, 620–5.

Nelson, P. V. & K. H. Hsiek 1971. Ammonium toxicity in chrysanthemum: critical level and symptoms. *Comm. in Soil Sci. and Plant Analysis* **2**, 439–48.

Nichols, D. G. 1981. Toxic pine bark – in potting mixes. *Seed and Nursery Trader*, May, 28–30.

Nichols, L. P. & M. H. Jordon 1972. Chemical soaks for prevention of growth of pathogenic organisms on clay and plastic pots. *Penn. State Flo. Gro. Bull.* **250**, 1 & 6–7.

Niemanand, R. H. & R. A. Clark 1976. Interactive effects of salinity and phosphorus nutrition on the concentrations of phosphate and phosphate esters in mature photosynthesizing corn leaves. *Plant Physiol.* **57**, 157–61.

North, C. P. & A. Wallace 1959. Nitrogen effects on chlorosis in macadamia. *Calif. Macadamia Soc. Yearbook*, **5**, 54–67.

Oertli, J. J. & O. R. Lunt 1962. Controlled release of fertiliser materials by incapsulating membranes: I Factors influencing the rate of release. *Soil Sci. Soc. Am. Proc.* **26**, 579–87.

Ogg, W. G. 1939. Peat. *Chemistry and Industry* **58**, 375–9.

Olsen, S. R. 1953. Inorganic phosphorus in alkaline and calcareous soils. *Agronomy* **4**, 89–122.

Owen, O. 1948. *The occurrence and correction of magnesium deficiency in* Solanum capsicastrum *under commercial conditions*. Rep. Expl Res. Stn Cheshunt, 80–1.

Parr, J. F. & A. G. Norman 1964. Effects of nonionic surfactants on root growth and cation uptake. *Plant Physiol.* **39**, 502–7.

Patterson, J. B. E. 1971. Metal toxicities arising from industry. In *Trace elements in soils and crops*. Tech. Bull. no. 21, 193–207. London: HMSO.

Paul, J. L. & C. I. Lee 1976. Relation between growth of chrysanthemums and aeration of various container media. *J. Am. Soc. Hort. Sci.* **101**, 500–3.

Paul, J. L. & A. T. Leiser 1968. Influence of calcium saturation of sphagnum peat on the rooting of fine woody species. *Hort. Res.* **8**, 41–50.

Paul, J. L. & E. Polle 1964. Nitrite accumulation related to lettuce growth in a slightly alkaline soil. *Soil Sci.* **100**, 292–7.

Paul, J. L. & L. V. Smith 1966. Rooting of chrysanthemum cuttings in peat as influenced by calcium. *Proc. Am. Soc. Hort. Sci.* **89**, 626–30.

Paul, J. L. & W. H. Thornhill 1969. Effects of magnesium on rooting of chrysanthemum. *J. Am. Soc. Hort. Sci.* **94**, 280–2.

Pearson, H. E. 1949. Effect of waters of different quality on some ornamental plants. *Proc. Am. Soc. Hort. Sci.* **53**, 532–42.

Penningsfeld, F. 1962. *Die Ernährung Im Blumen- und Zierplanzenbau*. Berlin/Hamburg: Paul Parey.

Peterson, J. C. 1981. Modify your pH perspective. *Florists' Review* **169**(4386), 34–5, 92 & 94.

Pickering, S. U. 1908. Action of heat and antiseptics on soils. *J. Agric. Sci.* **3**, 32–54.

Pokorny, F. A. 1979. Pine bark container media – an overview. *Combined Proc. Int. Plant Prop. Soc.* **29**, 484–95.

Pokorny, F. A. & H. Y. Wetzstein 1984. Internal porosity, water availability and root penetration of pine bark particles. *HortScience* **19**, 447–9.

Pokorny, F. A., H. A. Mills & P. Napier 1977. Determination of nitrate in a pine bark substrate. *Comm. in Soil Sci. and Plant Analysis* **8**, 87–95.

Poole, R. T. & C. A. Conover 1979. Melaleuca bark and solite as potential potting ingredients for foliage plants. *Ploc. Fla State Hort. Soc.* **92**, 327–9.

Poole, R. T. & R. W. Henley 1981. Constant fertilisation of foliage plants *J. Am. Soc. Hort. Sci.* **106**, 61–3.

Poole, R. T., C. A. Conover & J. N. Joiner 1976. Chemical composition of good quality tropical foliage plants. *Proc. Fla State Hort. Soc.* **89**, 307–8.

Powell, R. 1968. *Controlled release fertilisers*, p. 277. Noyes Development Corporation, Park Ridge, New Jersey, USA.

Prasad, M. 1983. *Container mixes for pot plants and nurseries*. Aglink 4/3000/1/83; HPP138 Pubs Ministry Agriculture and Fisheries, New Zealand.

Prasad, M. & P. A. Gallagher 1972. Sulphur coated urea, casein and other slow release nitrogen fertilizers for tomato production. *Acta Horticulturae* **26**, 165–73.

Puustjärvi, V. 1968. Cation exchange capacity in sphagnum mosses and its effect on nutrient and water absorption. *Peat and Plant News* **4**, 54–8.

Puustjärvi, V. 1969. Basin-Peat culture. *Peat and Plant News* **2**, 20–4.

Puustjärvi, V. 1970a. Mobilisation of nitrogen in peat culture. *Peat and Plant News* **3**, 35–42.

Puustjärvi, V. 1970b. Degree of decomposition. *Peat and Plant News* **4**, 48–52.

Puustjärvi, V. 1983. Nature of changes in peat properties during decomposition. *Peat and Plant Yearbook 1981–2*, 5–20.

Rader, L. F., L. M. White & C. W. Wittaker 1943. The salt index – a measure of the effect of fertilisers on the concentration of the soil solution. *Soil Sci.* **55**, 201–18.

Rattink, H. 1982. Disinfection of potting soil by means of gamma-radiation. *Med. Fac. Landbouww. Rijksuniv. Gent* **47**, 869–73.

Read, P. E. & R. Sheldrake, Jr 1966. Correction of chlorosis in plants grown in Cornell Peat-Lite mixes. *Proc. Am. Soc. Hort. Sci.* **88**, 576–81.

Reeker, R. 1960. The prevention of molybdenum deficiency in growth media from peat. *Soils and Fertilisers* **23**, abstract no. 2869.

Richards, D., M. Lane & D. V. Beardsell 1986. The influence of particle-size distribution in pinebark:sand:brown coal potting mixes on water supply, aeration and plant growth. *Scientia Hort.* **29**, 1–14.

Richards, L. A. (ed.) 1954. *Diagnosis and improvement of saline and alkaline soils.* United States Dept. of Agriculture, Agricultural Handbook no. 60.

Richards, M. 1974. *Drainage from containers*, p. 8. Massey University, New Zealand.

Roll-Hansen, J. 1975. Fritted trace elements (F.T.E.) as a basic fertilizer for peat. *Acta Horticulturae* **50**, 119–24.

Roorda van Eyzinga, J. P. N. L. & K. W. Smilde 1981. *Nutritional disorders in glasshouse tomatoes, cucumbers and lettuce.* Wageningen: PUDOC.

Rose, D. A. 1973. Some aspects of the hydrodynamic dispersion of solutes in porous materials. *J. Soil Sci.* **24**, 284–95.

Saynor, M. & D. M. Ann 1983. Mixing pesticides to protect block plants. *Growr*, 27 Oct., 43–6.

Schreiner, O. & E. C. Shorey 1909. *The isolation of harmful organic substances from soils.* USDA Bureau of Soils Bull. 53.

Scott, M. A. 1982. *The role of phosphate in the production of quality container-grown nursery stock.* Leaflet no. 2, Efford Expl Hort. Stn. London: HMSO.

Scott, M. A. 1984. *Hardy nursery stock, the use of bark in composts – A review of work at Efford EHS 1981–3.* Rev. Efford Expl Hort. Stn for 1983, 23–34.

Scott, M. A. 1985. *Annual Review 1984*, Efford Expl Hort. Stn, 15–22.

Scott, M. A., P. M. Smith & J. Evans 1984. Clean and clear, how to keep nursery irrigation water disease free. *Gardeners' Chronicle and Horticultural Trade J.* **195**(25), 12–14.

Seeley, J. G. 1947. Automatic watering of potted plants. *New York State Flo. Gro. Bull.* **23**, 1–9.

Sheard, G. F. 1940. *An investigation of the partial sterilisation of soil for horticultural purposes with special reference to the use of electrical sterilising equipment.* MSc. Thesis, University of Leeds.

Sheldrake, R., Jr & O. A. Matkin 1971. Wetting agents for peat moss. *Acta Horticulturae* **18**, 37–42.

Sheldrake, R., G. E. Doss, L. E. St John, Jr & D. J. Lisk 1978. Lime and charcoal amendments reduce fluoride absorption by plants cultured in a perlite–peat medium. *J. Am. Soc. Hort. Sci.* **103**, 268–70.

Shibata, A., T. Fujita & S. Maeda 1980. Nutricote. Coated fertilisers processed with polyolefin resins. *Acta Horticulturae* **99**, 179–86.

Smilde, K. W. 1971. *Evaluation of fritted trace elements on peat substrates.* Institute for Soil Fertility, Groningen, Rapport 1.

Smiley, R. W. 1974. Rhizosphere pH as influenced by plants, soils and nitrogen fertilisers. *Soil Sci. Soc. Am. Proc.* **38**, 795–9.

Smith, I. E. 1984. *Nutrition of seedlings grown in soilless media in compartmentalised trays.* Report Dept Hort. Sci., Univ. of Natal, South Africa.

Smith, J. H. 1964. Relationships between soil cation-exchange capacity and the toxicity of ammonia to the nitrification process. *Soil Sci. Soc. Am. Proc.* **28**, 640–4.

Soil and Plant Laboratory, California 1982. Handbook.

Solbraa, K. 1974. Different bark qualities and their uses in plant cultivation. *Proc. Symp. on the standardisation of bark compost in horticulture*, Ghent, 78–85.

Solbraa, K. 1979a. Composting of bark. I Different bark qualities and their uses in plant production. *Reports of the Norwegian Forest Res. Inst.* **34**, 13.

Solbraa, K. 1979b. Composting of bark. IV Potential growth-reducing compounds and elements in bark. *Reports of Norwegian Forest Res. Inst.* **34**, 16.

Solbraa, K. 1979c. Composting of bark. III Experiments on a semi-practical scale. *Reports of the Norwegian Forest Res. Inst.* **34**, 15.

Solbraa, K. & A. R. Selmer-Olsen 1981. Manganese toxicity – in particular when growing plants in bark compost. *Acta Agriculturae Scandinavica* **31**, 29–39.

Sonneveld, C. 1979. Changes in chemical properties of soil caused by steam sterilization. In *Soil disinfestation*, D. Mulder (ed.), 39–50. Amsterdam: Elsevier Scientific.

Sonneveld, C. & J. van Beusekom 1974. The effect of saline irrigation water on some vegetables under glass. *Acta Horticulturae* **35**, 75–85.

Sonneveld, C. & S. Voogt 1973. The effects of steam sterilization with steam-air mixtures on the development of some glasshouse crops. *Plant and Soil* **38**, 415–23.

Sonneveld, C. & W. Voogt 1983. Studies on the salt tolerance of some flower crops grown under glass. *Plant and Soil* **74**, 41–52.

Sonneveld, C., J. van den Ende & P. A. van Dijk 1974. Analysis of growing media by means of a 1 : 1½ volume extract. *Comm. in Soil Sci. and Plant Analysis* **5**, 183–202.

Spomer, L. A. 1974. Optimizing container soil amendment: the 'threshold proportion' and prediction of porosity. *HortScience* **9**, 532–3.

Spomer, L. A. 1975a. Small soil containers as experimental tools: soil water relations. *Comm. in Soil Sci. and Plant Analysis* **6**, 21–6.

Spomer, L. A. 1975b. Availability of water absorbed by hardwood bark soil amendment. *Agronomy J.* **67**, 589–90.

Spomer, L. A. & R. W. Langhans 1975. The growth of greenhouse bench *Chrysanthemum morifolium* at high soil water and aeration. *Comm. in Soil Sci. and Plant Analysis* **6**, 545–53.

Stanek, W. & I. A. Worley 1984. A terminology of virgin peat and peatlands *Proc. Int. Symp. on Peat Utilization*, Oct. 1983, Bemidji State Univ., USA, 75–102.

Sterrett, S. B., C. W. Reynolds, F. D. Schales, R. L. Chaney & L. W. Douglass 1982. Transplant quality and metal concentrations in vegetable transplants grown in media containing sewage sludge compost. *HortScience* **17**, 920–2.

Still, S. M. 1976. Growth of 'Sunny Mandalay' chrysanthemums in hardwood-bark-amended media as affected by insolubilized poly(ethylene oxide). *HortScience* **11**, 483–4.

Still, S. M., M. A. Dirr & J. B. Gartner 1976. Phytotoxic effects of several bark extracts on mung bean and cucumber growth. *J. Am. Soc. Hort. Sci.* **101**, 34–7.

Swaine, D. J. 1962. *Trace element content of fertilisers*. Commonwealth Bureau of Soils Technical Communication, no. 52.

Tew Schrock, P. A. & K. L. Goldsberry 1982. Growth responses of seed geranium and petunia to N source and growing media. *J. Am. Soc. Hort. Sci.* **107**, 348–52.

Thomas, M. B. 1980. Phosphorus response of *Proteaceae* and other nursery plants in containers. *R. N.Z. Inst. Hort., Ann. J.* **8**, 21–33.

Thompson, L. M. 1957. *Soils and fertility*. New York: McGraw-Hill.

Tod, H. 1956. High calcium or high pH? A study of the effect of soil alkalinity on the growth of *Rhododendron. J. Scottish Rock Garden Club*, **V**, 1–8.

Troug, E. 1948. Lime in relation to availability of plant nutrients. *Soil Sci.* **65**, 1–7.

Tschabold, E. E., W. C. Meredith, L. R. Guse & E. V. Krumkans 1975. Ancymidol performance as altered by potting media composition. *J. Am. Soc. Hort. Sci.* **100**, 142–4.

Valoras, N., J. Letey, J. P. Martin & J. F. Osborn 1976. Degradation of a nonionic surfactant in soils and peat. *Soil Sci. Soc. Am. J.* **40**, 60–3.

Verdonck, O., D. de Vleeschauwer & R. Pennick 1983. Cocofibre dust, a new growing medium for plants in the tropics. *Acta Horticulturae* **133**, 215–20.

Verdure, M. 1981. Improvement of physical properties of black peat. *Acta Horticulturae* **126**, 131–42.

Verwer, F. L. J. A. W. 1978. Efficiency of capillary mattings by growing plants. *Acta Horticulturae* **82**, 231–43.

Vlamis, J. & R. D. Raabe 1985. Copper deficiency of manzanita grown in a bark-sand mixture. *HortScience* **20**, 61–2.

Wadleigh, C. H. & A. D. Ayers 1945. Growth and biochemical composition of bean plants as conditioned by soil moisture tension and salt concentration. *Plant Physiol.* **20**, 106–32.

Walker, T. W. & R. Thompson 1949. Some observations on the chemical changes effected by the steam sterilisation of glasshouse soils. *J. Hort. Sci.* **25**, 19–35.

Wall, R. F. & F. B. Cross 1943. *Greenhouse studies on the toxicities of Oklahoma salt contaminated waters*. Okla. Agric. Expl Stn Tech. Bull., T-20.

Wallace, A. & O. R. Lunt 1960. Iron chlorosis in horticultural plants. A review. *Proc. Am. Soc. Hort. Sci.* **75**, 819–41.

Walsh, T. & T. A. Barry 1958. The chemical composition of some Irish peats. *Proc. R. Irish Academy* **59B**, 305–28.

van Wambeke, E., J. Wijsmans & P. D'Hertefelt 1983. Possibilities in microwave application for growing substrate disinfestation. *Acta Horticulturae* **152**, 209–13.

Warncke, D. D. & D. M. Krauskopf 1983. *Greenhouse growth media: testing and nutrition guidelines*. Extension Bull. E-1736, Michigan State University.

Warren Wilson, J. & J. Tunny 1965. Defects of perlite as a medium for plant growth. *Aust. J. Expl Agric. and Animal Husbandry* **5**, 137–40.

Waters, W. E., W. Llewellyn & J. Nesmith 1970. The chemical, physical and salinity characteristics of twenty-seven soil media. *Proc. Fla State Hort. Soc.* **83**, 482–8.

Waters, W. E., J. Nesmith, C. M. Geraldson S. S. Woltz 1972. *The interpretation of soluble salt tests and soil analysis by different procedures*. Bradenton, Florida, AREC Mimeo Report GC-1972-4.

Watson, W. 1913. Soils suitable for azalea culture. In *The gardener's assistant*, vol. 1, 150–4. London: Gresham Publishing.

Weatherspoon, D. M. & C. C. Harrell 1980. Evaluation of drip irrigation for container production of woody landscape plants. *HortScience* **15**, 488–9.

Wells, D. A. & R. Soffe 1962. A bench method for the automatic watering by capillarity of plants grown in pots. *J. Agric. Engng Res.* **7**, 42–6.

Wheeler, R. M., S. H. Schwartzkopf, T. W. Tibbitts & R. W. Langhans 1985. Elimination of toxicity from polyurethane foam plugs used for plant culture. *HortScience* **20**, 448–9.

White, J. W. 1971. Interaction of nitrogenous fertilizers and steam on soil chemicals and carnation growth. *J. Am. Soc. Hort. Sci.* **96**, 134–7.

White, J. W. 1974. Dillon Research Fund, progress report on research at Penn. State. *Penn. Flo. Gro. Bull.* **273**, 3–4.

White, J. W. & J. W. Mastalerz 1966. Soil moisture as related to container capacity. *Proc. Am. Soc. Hort. Sci.* **89**, 758–65.

Whitt, D. M. & L. D. Baver 1930. Particle size in relation to base exchange capacity and hydration properties of Putnam clay. *J. Am. Soc. Agron.* **29**, 703–8.

Wilcox, G. E., J. E. Hoff & C. M. Jones 1973. Ammonium reduction of calcium and magnesium content of tomato and sweet corn leaf tissue and influence on blossom end rot of tomato fruit. *J. Am. Soc. Hort. Sci.* **98**, 86–9.

Wilkerson, D. C. 1981. *Some physical and chemical properties of pine bark growing medium used as an evaluation of its nutrition status*. PhD Thesis, Dept of Horticulture, Louisiana State University.

Wilkerson, D. C. & E. N. O'Rourke 1983. The effects of three analytical systems on the interpretation of nutrient 'availability' in pink bark growing media. *HortScience* **18**, 301–2.

Wind, G. P. 1968. *Root growth in acid soils*. The Netherlands Inst. for Land and Water Management Research, Tech. Bull. 55.

Winsor, G. W. & M. I. E. Long 1956. Mineralisation of the nitrogen in urea-formaldehyde compounds in relation to soil pH. *J. Sci. Food Agric.* **7**, 560–4.

Winsor, G. W., J. N. Davies & D. M. Massey 1963. Salinity studies. I Effect of calcium sulphate on the correlation between plant growth and electrical conductivity of soil extracts. *J. Sci. Food Agric.* **14**, 42–8.

Worrall, R. J. 1978. The use of composted wood waste as a peat substitute. *Acta Horticulturae* **82**, 81–6.

Worrall, R. J. 1981a. Comparison of composted hardwood and peat-based media for the production of seedlings, foliage and flowering plants. *Scientia Hort.* **15**, 311–19.

Worrall, R. J. 1981b. High temperature release characteristics of resin-coated slow release fertilisers. *Combined Proc. Int. Plant Prop. Soc.* **31**, 176–81.

Yazaki, Y. & D. Nichols 1978. Phytotoxic components of *Pinus radiata* bark. *Aust. For. Res.* **8**, 185–98.

Yeager, T. H. & R. D. Wright 1982. Phosphorus requirement of *Ilex crenata* Thumb. cv. Helleri grown in a pine bark medium. *J. Am. Soc. Hort. Sci.* **107**, 558–62.

Index

acidification of media 84–6
  of water 222, 225–7
acidity (*see also* pH) 79–86
  bark 24–5
  effect of fertilizers 195, 221
  peat 9, 19, 81–5
acids and water quality 225–7
air-filled porosity 40, 48–50, 55
  and container depth 46
  oxygen diffusion rate 51–3
  management 50
  methods of determination 50
  requirement of plants 49
  time integration 49
Aldrin 270
algaecides, use in irrigation systems 230, 232
aluminium 163
  availability and pH 163
  hydrangea flower colour 163
  phosphate 67, 122
  sulphate, rate of use 163
  toxicity 263
ammonia, free 29, 101–2, 260
  toxicity 102–3
ammonifying bacteria 101, 105–7, 241–2, 260
ammonium, fixation by vermiculite 34, 71
  pH effect 80, 99, 106
  toxicity 99–100, 112
ammonium polyphosphates 125, 217
anaerobic toxins 53
analytical methods 68, 71–6, 80
animal fibre 30
anions 67–8
  exchange capacity 34, 38
antagonism between elements 77, 99, 110, 121, 135, 161
aquatic plant compost 28
available water 40, 44, 53, 56–9, 62
azalea mixes 192–6

bacteria: ammonifying 101, 105–7, 241–2, 260
  nitrifying 101, 103, 105–7, 241
bactericides in irrigation systems 230, 232
bark 21–8, 187, 189, 192
  ageing 26

bulk density 22, 42
  cation exchange capacity 24
  chemical characteristics 22–5
  composting 26–8
  decomposition rate 23
  fungal suppression 28
  hardwood 24–5
  lime requirement 28
  manganese toxicity 26, 28
  mineral composition 22
  nitrogen requirement 23–4, 27–8, 189
  pH 24–5
  physical characteristics 21–2, 24
  phytotoxicities 25–7
  softwood 21–4
bench type, effect on container drainage 233, 273
bicarbonates 190, 224–7
  effect on media pH 221
  neutralization with acids 222, 225
blockages, irrigation systems 230
blossom-end rot, tomatoes 133
boiler feed water, treatment 253
boron 153–6
  availability and pH 81, 112, 153
  content of fertilizers 155, 170, 286
  deficiency and nitrogen source 112, 153, 165
  deficiency levels in foliage 171
  deficiency symptoms 153
  foliar sprays 173
  fritted form 155, 163–5
  in liquid feeds 156
  requirement in mixes 155
  toxicity 155, 171–2
British Thermal Unit 250
bulk density 40, 42
  bark 22
  peat 13, 19
  peat-sand mixtures 56, 187
  perlite 39
  plastics 39
  minerals 32–3
  vermiculite 39

(C–A) balance 98, 194
cadmium toxicity 30

301